Characterization of Nanomaterials

Characterization of Nanomaterials

Selected Papers from 6th Dresden Nanoanalysis Symposium

Editors

Ehrenfried Zschech
Robert Sinclair
Rodrigo Martins
Marco Sebastiani
Sabrina Sartori

MDPI • Basel • Beijing • Wuhan • Barcelona • Belgrade • Manchester • Tokyo • Cluj • Tianjin

Editors
Ehrenfried Zschech
Fraunhofer Institute for Ceramic Technologies and Systems, Microelectronic Materials and Nanoanalysis
Germany

Robert Sinclair
Stanford University
USA

Rodrigo Martins
Universidade NOVA de Lisboa and CEMOP/UNINOVA
Portugal

Marco Sebastiani
Università degli studi Roma Tre
Italy

Sabrina Sartori
University of Oslo
Norway

Editorial Office
MDPI
St. Alban-Anlage 66
4052 Basel, Switzerland

This is a reprint of articles from the Special Issue published online in the open access journal *Nanomaterials* (ISSN 2079-4991) (available at: https://www.mdpi.com/journal/nanomaterials/special_issues/charact_nano).

For citation purposes, cite each article independently as indicated on the article page online and as indicated below:

LastName, A.A.; LastName, B.B.; LastName, C.C. Article Title. *Journal Name* **Year**, *Volume Number*, Page Range.

ISBN 978-3-0365-0756-9 (Hbk)
ISBN 978-3-0365-0757-6 (PDF)

Cover image courtesy of Marco Sebastiani.

Contents

About the Editors

Ehrenfried Zschech is Department Head for Microelectronic Materials and Nanoanalysis at the Fraunhofer Institute for Ceramic Technologies and Systems in Dresden, Germany, which he joined in 2009. His responsibilities include multi-scale materials characterization and reliability engineering. Ehrenfried Zschech received his Dr. rer. nat. degree from Technische Universität Dresden. After having spent four years as a project leader in the field of metal physics and reliability of microelectronic interconnects at Research Institute for Nonferrous Metals in Freiberg, he was appointed as a university teacher for ceramic materials at Freiberg University of Technology. In 1992, he joined the development department at Airbus in Bremen, where he managed the metal physics group and studied the laser-welding metallurgy of aluminum alloys. From 1997 to 2009, Ehrenfried Zschech managed the Materials Analysis Department and the Center for Complex Analysis at Advanced Micro Devices in Dresden. In this position, he was responsible for the analytical support for process control and technology development in leading-edge semiconductor manufacturing, as well as for physical failure analysis. He holds an adjunct professorship at Faculty of Chemistry of Warsaw University, Poland, as well as honorary professorships for Nanomaterials at Brandenburg University of Technology Cottbus and for Nanoanalysis at Technische Universität Dresden. He has published three books and he has authored or co-authored more than 200 papers in peer-reviewed journals in the areas of materials science, solid-state physics and reliability engineering. Ehrenfried Zschech is a Member of the Board of Directors of the Materials Research Society (MRS), a Member of the Senate of the European Materials Research Society (E-MRS) and an Honorary Member of the Federation of the European Materials Societies (FEMS). In 2019, he was awarded the FEMS European Materials Gold Medal.

Robert Sinclair is the Charles M. Pigott Professor in the School of Engineering, in the Department of Materials Science and Engineering at Stanford University. He received his PhD in Materials Science from Cambridge University in 1972. He first worked as a Research Engineer at the University of California, Berkeley from 1973 to 1977, before joining the faculty of Stanford University in the Department of Materials Science and Engineering in 1977. He was the Department Chair from 2004 to 2014, and served as the Director of Stanford Nanocharacterization Laboratory from 2002 to 2013, as well as the Director of Bing Overseas Studies Program from 2010 to 2012. He is currently Charles M. Pigott Professor in the School of Engineering, and is the Director of the Wallenberg Research Link. Professor Sinclair's research interests include development and application of advanced transmission electron microscopy techniques, including aberration correction and in situ methods, studies of material and thin film reactions in energy-related materials, semiconductor processing, magnetic computer hard discs and nanomaterials for cancer detection. He was Chair of the National Research Council Committee on Smaller Facilities, the findings of which were published by the National Academy of Science in 2006 ("Mid-size Facilities: The Infrastucture of Materials Research"). Among his recent awards and honors are the Festschrift in honor of Robert Sinclair and Nestor Zaluzec, PICO 2017 Workshop (Germany); the John M. Cowley Distinguished Lecturer, at Arizona State University in 2015; the David Turnbull Lectureship of the Materials Research Society in 2012; the Distinguished Scientist Award (Physical Sciences), Microscopy Society of America in 2009.

Rodrigo Martins, President of the European Academy of Sciences; President of the International Union of Materials Research Societies; Full Professor at FCT-NOVA. Member of the:

• Scientific Council of the European Research Council
• Portuguese Academy of Engineering
• Portuguese Order of Engineers, OE
• Board of Admission and Qualification of OE.

Rodrigo Martins is the founder and director of the Centre of Excellence in Microelectronics and Optoelectronics Processes of Uninova; leader of the Materials, Optoelectronics and Nanotechnologies group of I3N/CENIMAT and its sub-director; member of the nomination committee of the EIT KIC Raw Materials, Editor in Chief of the journal *Discover Materials*. He is an expert in the field of advanced functional materials, nanotechnologies, microelectronics, transparent electronics (pioneer) and paper electronics (inventor), with more than 575 papers published in WoK; Member of the:

• Steering Committee of European Technology Platform for Advanced Engineering Materials and Technologies, EuMat.
• Joint Innovation Centre for Advanced Material Sino-Portuguese.
• Administration board of the nature journal: npj 2D Materials and Applications.

Rodrigo Martins was decorated with the gold medal of merit and distinction by the Almada Municipality for his R&D achievements, in 2016. He received more than 18 international and national prizes and distinctions for his scientific work. ORCID: http://orcid.org/0000-0002-1997-7669: Webpage: https://cemop.uninova.pt/.

Marco Sebastiani is currently an associate professor of Materials Science at Roma Tre University, where he is also a lecturer of materials science and technology for aeronautical engineering. Currently, his major research efforts are in the fields of surface engineering, micro-device development, thin films and small-scale mechanical behaviour. The specific focus is on (a) Micro-scale residual stress analysis by a focused ion beam (FIB) approach, (b) Nano-mechanical characterization of advanced materials and devices using nanoindentation testing, (c) Nanostructured thin film modelling and production by physical vapor deposition (PVD) processes. He is one the editors of the international peer-reviewed journal *Materials & Design* (IF 6.3), reviewer for a wide range of international journals, and participant in European and national R&D projects (among which, the role of coordinator of the FP7 and H2020 large collaborative projects: iSTRESS, www.istress.eu and OYSTER, www.oyster-project.eu). Since 2014, he is actively involved in the most important European actions for promoting the access to advanced materials characterization tools for the manufacturing industry. In particular, he is a member of the Operational Management Board at the European Materials Characterization Council (EMCC, www.characterisation.eu). He is the co-author of more than 95 papers in ISI journals, with over 2300 citations and an h-index of 28 (http://orcid.org/0000-0002-9574-1578).

Sabrina Sartori is an associate professor at the Department of Technology Systems at University of Oslo. She studied physical chemistry at the University of Padova and graduated from University of Bologna in 2003 with a PhD degree in Materials Science and Engineering. She joined the University of Oslo in 2013, after working as a research scientist at the Institute for Energy Technology in Norway. Her research interests and expertise cover solid state physics and chemistry, including the synthesis of materials for hydrogen storage and batteries and their structure characterization via synchrotron- and neutron-based methods at large-scale facilities. Sartori established and is leading the laboratory of solid-state synthesis, and the laboratory of energy storage systems at UiO, and is the responsible of the two years Master of science program "Renewable Energy Systems". She serves as an expert and leader in various committees, boards and initiatives, including at the International Energy Agency (IEA)—Hydrogen Implementing Agreement. She received several honors for her research activity in energy storage, for instance, the Feinberg Foundation visiting faculty program award. She organized several national and international events, for instance, as co-Chair of MRS Fall meeting 2020 in Boston.

nanomaterials

MDPI

Editorial

Editorial for the Special Issue "Characterization of Nanomaterials: Selected Papers from 6th Dresden Nanoanalysis Symposium"

Ehrenfried Zschech [1,*], Robert Sinclair [2], Rodrigo Martins [3], Marco Sebastiani [4] and Sabrina Sartori [5]

1 Fraunhofer Institute for Ceramic Technologies and Systems (IKTS), Microelectronic Materials and Nanoanalysis, 01109 Dresden, Germany

2 Department of Materials Science and Engineering, Stanford University, Stanford, CA 94305, USA; bobsinc@stanford.edu

3 i3N/CENIMAT, Department of Materials Science, Faculty of Science and Technology, Universidade NOVA de Lisboa and CEMOP/UNINOVA, Campus de Caparica, 2829-516 Caparica, Portugal; rfpm@fct.unl.pt

4 Engineering Department, Università degli studi Roma Tre, via dellaVasca Navale, 79, 00146 Rome, Italy; seba@uniroma3.it

5 Department of Technology Systems, University of Oslo, NO-2027 Kjeller, Norway; sabrina.sartori@its.uio.no

* Correspondence: ehrenfried.zschech@ikts.fraunhofer.de; Tel.: +49-172-252-34-70

Received: 10 October 2019; Accepted: 18 October 2019; Published: 27 October 2019

More than ever before, materials-driven product innovations in industry and shorter time-to-market introductions for new products require high advancement rates and a tight coupling between research, development and manufacturing. This approach, where scientists and engineers from industry and research institutes work together, includes sustained progress in materials science and engineering, and in materials and process characterization. Analytical techniques and respective tools, particularly to investigate nanomaterials, are considered to be fundamental drivers for innovation in industry.

This Special Issue "Characterization of Nanomaterials" collects nine selected papers presented at the 6th Dresden Nanoanalysis Symposium, held at Fraunhofer Institute for Ceramic Technologies and Systems in Dresden, Germany, on 31 August 2018. The Dresden Nanoanalysis Symposium was organized by the Dresden Fraunhofer Cluster Nanoanalysis (DFCNA), supported by the European Materials Research Society (E-MRS) and the European Materials Characterization Council (EMCC). Following the specific motto of this annual symposium "Materials challenges—Micro- and nanoscale characterization", it covered various topics of nanoscale materials characterization along the whole value and innovation chain, from fundamental research up to industrial applications. It brought about 100 scientists and engineers together from universities, research institutions, equipment manufacturers, and industrial end-users. New results in disruptive nanoanalysis techniques were reported in several talks and in the poster sessions, and novel solutions in the field of nanoscale materials characterization for process and quality control were shown.

The scope of this Special Issue is to provide an overview of the current status, recent developments and research activities in the field of nanoscale materials characterization, with a particular emphasis on future scenarios. Primarily, analytical techniques for the characterization of thin films and nanostructures [1–9] are discussed, including modeling and simulation [9]. Particular techniques for materials characterization are 3D time-of-flight secondary ion mass spectroscopy [1], in situ X-ray diffraction [6] and electron backscatter diffraction [8]. Several papers cover advanced nanostructured materials for future electronic devices and sensors [2,3,5,7]. In addition, degradation processes in battery electrodes [4] and the growth kinetics of intermetallic phases [8] are addressed. We anticipate this

Special Issue to be accessible to a wide audience, as it explores not only methodical aspects of nanoscale materials characterization, but also materials synthesis, fabrication of devices and applications.

The first paper describes the application of time-of-flight secondary ion mass spectroscopy (ToF-SIMS) in 3D microstructures, specifically so-called through-silicon vias, i.e., high-aspect ratio metal contacts needed for connecting 3D-stacked microchips [1]. This novel methodical approach provides several advantages compared to other frequently used techniques such as electron microscopy. The authors used data image slicing of the 3D ToF-SIMS analysis to study the uniformity of the silicon dopant concentration in atomic layer deposited (ALD) HfO_2 thin films.

Barros et al. report on the role of structure and composition on the performances of p-type SnO_x thin-field transistors (TFTs) with a bottom gate configuration deposited by radio frequency magnetron sputtering at room temperature, followed by a post-annealing step [2]. X-ray diffraction (XRD) and Mössbauer spectroscopy allow the authors to identify the best phases/compositions and thicknesses (around 12 nm) to fabricate p-type TFTs. Moreover, the authors provide an overview that presents latest developments in SnO_x TFT processing.

Studying multi-level resistive switching characteristics of a Cu_2O/Al_2O_3 bilayer device, other authors found that an oxidation state gradient in copper oxide induced by the fabrication process plays a dominant role in defining multiple resistance states [3]. The highly conductive grain boundaries of copper oxide—an unusual property for an oxide semiconductor—are discussed for the first time regarding their role in the resistive switching mechanism.

The authors of paper [4] study materials aging in nickel manganese cobalt (NMC) oxides and related lifetime-reducing degradation processes in cathodes of lithium-ion batteries. Sieber et al. present an approach to recover NMC particles from lithium-ion battery cathodes while preserving their chemical and morphological properties, with a minimal use of chemicals. The key task was the separation of the cathode coating layer consisting of NMC, an organic binder, and carbon black, from the Al substrate foil. This can be performed in water under strong agitation to support the slow detachment process. The authors mitigate negative effects such as dissolving the Al substrate foil and $Al(OH)_3$ precipitation using pH-adjusted solutions with sufficiently high buffer capacities.

Dobosz et al. study the oxide layer formed on the surface of the Ga–Sn–Zn eutectic alloy using atomic force microscopy (AFM), X-ray photoelectron spectroscopy (XPS) and transmission electron microscopy (TEM) [5]. The authors find that it is possible to obtain nanocrystalline oxide layers that contain about 90 at-% gallium with some additions of tin and zinc.

Another paper describes the synthesis of upconverter nanostructures composed of an yttrium oxide host matrix co-doped with ytterbium and europium, i.e., $Y_2O_3:Yb^{3+}/Eu^{3+}$ [6]. These nanostructures were characterized by X-ray diffraction (XRD), scanning transmission electron microscopy (STEM) and scanning electron microscopy (SEM). The acetic-based nanostructures result in nanosheets with a thickness of about 50 nm, while hydrochloric and nitric-based ones result in sphere-shaped nanostructures.

Carvalho et al. report about printed and flexible inorganic electrolyte-gated transistors (EGTs) on paper, with a channel layer based on interconnected zinc oxide (ZnO) nanoparticles [7]. The ZnO nanoparticles were dispersed by ethyl cellulose (EC), an eco-friendly binder. Fully printed devices on glass substrates using a composite solid polymer electrolyte as gate dielectrics exhibit a saturation mobility above 5 $cm^2\ V^{-1}\ s^{-1}$ after annealing at 350 °C. The authors optimize the nanoparticle content in the ink, resulting in the formation of a ZnO channel layer at a maximum annealing temperature of 150 °C, which is compatible with paper substrates.

The reactivity and kinetics of intermetallic phase growth in the Ni/Al/Ni system for nickel substrates while varying size and shape of the Ni grains are studied by Kwiecien et al. [8]. The sequence of the formation of particular intermetallic phases is determined using scanning electron microscopy (SEM), energy dispersive X-ray spectroscopy (EDX), and electron backscattered diffraction (EBSD) as well as transmission electron microscopy (TEM).

Finally, this Special Issue features a study of the influence of nanoscale residual stress depth gradients on the nanomechanical behavior and adhesion energy of aluminum nitride (AlN) and Al/AlN thin films sputtered on a (100) silicon substrate [9]. Using a focused ion beam (FIB) incremental ring-core method, the residual stress depth gradient is assessed in the films. The adhesion energy was then quantified using a nanoindentation-based model. The authors show that an additional Al bond layer and inhomogeneous residual stresses increase the tensile stress at the coating/substrate interface and, consequently, negatively affect the adhesion of AlN to a substrate such as silicon.

The papers collected here reflect the existing widespread interest in materials characterization for materials research, development, and innovation, but also for process and quality control in industry, and provide an insight particularly into the directions in which new developments in characterization techniques are currently headed. In this Special Issue, material transitions that are necessary to improve the performance and to maintain the reliability of products for applications in microelectronics and energy storage are highlighted.

As lively discussed and consistently found during the symposium, research and development in materials characterization techniques are increasingly needed for modern materials science, for innovation in high-tech branches and to guarantee the functionality, performance and reliability of advanced products. We hope that this Special Issue will stimulate fruitful discussions and co-operation between experts in academia and industry, who are working in the field of advanced materials characterization.

Author Contributions: All authors contributed to the editorial.

Funding: This research received no external funding.

Conflicts of Interest: The authors declare no conflict of interest.

References

1. Kia, A.M.; Haufe, N.; Esmaeili, S.; Mart, C.; Utriainen, M.; Puurunen, R.L.; Weinreich, W. ToF-SIMS 3D analysis of thin films deposited in high aspect ratio structures via atomic layer deposition and chemical vapor deposition. *Nanomaterials* **2019**, *9*, 1035. [CrossRef] [PubMed]
2. Barros, R.; Saji, K.J.; Waerenborgh, J.C.; Barquinha, P.; Pereira, L.; Carlos, E.; Martins, R.; Fortunato, E. Role of structure and composition on the performances of P-type tin oxide thin-film transistors processed at Low-Temperatures. *Nanomaterials* **2019**, *9*, 320. [CrossRef] [PubMed]
3. Deuermeier, J.; Kiazadeh, A.; Klein, A.; Martins, R.; Fortunato, E. Multi-Level Cell Properties of a Bilayer Cu_2O/Al_2O_3 Resistive Switching Device. *Nanomaterials* **2019**, *9*, 289. [CrossRef] [PubMed]
4. Sieber, T.; Ducke, J.; Rietig, A.; Langner, T.; Acker, J. Recovery of $Li(Ni_{0.33}Mn_{0.33}Co_{0.33})O_2$ from Lithium-Ion Battery Cathodes: Aspects of Degradation. *Nanomaterials* **2019**, *9*, 246. [CrossRef] [PubMed]
5. Dobosz, A.; Daeneke, T.; Zavabeti, A.; Zhang, B.Y.; Orrell-Trigg, R.; Kalantar-Zadeh, K.; Wójcik, A.; Maziarz, W.; Gancarz, T. Investigation of the Surface of Ga–Sn–Zn Eutectic Alloy by the Characterisation of Oxide Nanofilms Obtained by the Touch-Printing Method. *Nanomaterials* **2019**, *9*, 235. [CrossRef] [PubMed]
6. Nunes, D.; Pimentel, A.; Matias, M.; Freire, T.; Araújo, A.; Silva, F.; Gaspar, P.; Garcia, S.; Carvalho, P.A.; Fortunato, E.; et al. Tailoring Upconversion and Morphology of Yb/Eu Doped Y_2O_3 Nanostructures by Acid Composition Mediation. *Nanomaterials* **2019**, *9*, 234. [CrossRef] [PubMed]
7. Carvalho, J.T.; Dubceac, V.; Grey, P.; Cunha, I.; Fortunato, E.; Martins, R.; Clausner, A.; Zschech, E.; Pereira, L. Fully Printed Zinc Oxide Electrolyte-Gated Transistors on Paper. *Nanomaterials* **2019**, *9*, 169. [CrossRef] [PubMed]
8. Kwiecien, I.; Bobrowski, P.; Wierzbicka-Miernik, A.; Litynska-Dobrzynska, L.; Wojewoda-Budka, J. Growth Kinetics of the Selected Intermetallic Phases in Ni/Al/Ni System with Various Nickel Substrate Microstructure. *Nanomaterials* **2019**, *9*, 134. [CrossRef]

9. Ali, R.; Renzelli, M.; Khan, M.I.; Sebastiani, M.; Bemporad, E. Effects of Residual Stress Distribution on Interfacial Adhesion of Magnetron Sputtered AlN and AlN/Al Nanostructured Coatings on a (100) Silicon Substrate. *Nanomaterials* **2018**, *8*, 896. [CrossRef]

nanomaterials

MDPI

Article

Growth Kinetics of the Selected Intermetallic Phases in Ni/Al/Ni System with Various Nickel Substrate Microstructure

Izabella Kwiecien *, Piotr Bobrowski, Anna Wierzbicka-Miernik, Lidia Litynska-Dobrzynska and Joanna Wojewoda-Budka

Institute of Metallurgy and Materials Science, Polish Academy of Sciences, 30-059 Krakow, Poland; p.bobrowski@imim.pl (P.B.); a.wierzbicka@imim.pl (A.W.-M.); l.litynska@imim.pl (L.L.-D.); j.wojewoda@imim.pl (J.W.-B.)

* Correspondence: i.kwiecien@imim.pl; Tel.: +48-12-295-28-07

Received: 30 December 2018; Accepted: 20 January 2019; Published: 22 January 2019

Abstract: Reactivity in nickel–aluminum system was examined for two variants of nickel substrates in terms of the size and shape of Ni grains. The microstructure transformation aroused due to the annealing at 720 °C for different annealing times (0.25 to 72 h) was consequently followed. The sequence of formation of the particular intermetallic phases was given. The interconnection zones were examined by means of scanning electron microscopy supported with energy dispersive X-ray spectroscopy and electron backscattered diffraction techniques as well as by the transmission electron microscopy. The growth kinetics data for AlNi, $AlNi_{Ni\text{-}rich}$ and $AlNi_3$ phases for both variants of substrates was given, indicating the differences obtained in previous works on this subject.

Keywords: intermetallic phases; growth kinetics; Al–Ni system

1. Introduction

Nickel–aluminum system is well known and commonly researched reactive system. This is associated with the wide practical application of the $AlNi_3$ and AlNi intermetallic phases, which are used on a large scale in many industrial branches for example, as multilayers, oxidation resistant coatings (aluminizing treatment of nickel alloys), turbine blades (aircraft industry), electronics industry. To ensure the protection against the oxidation, aluminum-rich phases such as Al_3Ni, Al_3Ni_2 and AlNi are the most important ones, however, the greatest mechanical properties are associated with the presence of only two phases: AlNi and $AlNi_3$ [1–3].

β-AlNi and γ-$AlNi_3$ are cubic phases possessing B2 and L12 structure types, respectively. These intermetallics are characterized by high degree of order at elevated temperature and indicate high mechanical strength even at high temperature. Also they possess high thermodynamic stability in wide range of chemical composition and high degree of crystal lattice order with various amounts of defects. β-AlNi phase composition varies in the wide range from 40 to 55 at. % of Al at 700 °C. Additionally, within its structure many defects can be formed. Bradley and Taylor [4] have shown that nickel replaces aluminum in its lattice causing the excess of the nickel, however, in Ni-deficient compounds, aluminum does not replace the nickel in its lattice, causing formation of vacancies. It explains high number of structure's defects. As a result of high temperature diffusion processes, γ-$AlNi_3$ intermetallic phase can be created in the reaction between AlNi and Ni. Base structural Ni_3Al cell crystallizes in cubic lattice, where aluminum atoms are in the corners and nickel atoms are at the center of cell walls [4–6].

Many works consider the diffusion processes in Ni-Al system, however, as variables, only parameters such as annealing time and temperature of reaction either in solid state or solid/liquid state are analyzed. Additionally, the system is studied in many configurations with respect to the chemical composition of initial substrates (end members) and various their combinations [3,5,7–11]. Moreover, different experimental procedures have been employed in these studies such as: Ni/Al diffusion couples [3,5,8–10,12], Ni/Al/Ni sandwiches [7,13–15], or more complex assemblies such as multilayers, nanocoatings or aluminized nickel alloys [16–21].

One of the promising joining processes is the diffusion soldering (DS) [13,22–24], which is schematically presented in Figure 1. The main stage of this process is the isothermal solidification, where the reaction of liquid low melting interlayer (LM) and solid substrates (HM) takes place. With increasing of the DS temperature to the appropriate one—required in the process, the low melting component turns into the liquid state and the reaction at the solid/liquid interface (between the high and low melting components) begins. After some time, the first intermetallic phase is formed, then, in the next stage of DS diffusion in solid state between the obtained intermetallic phase and the high-melting substrate takes place. When the whole liquid metal is consumed, reaction proceeds only in the solid state. Phases are created and consumed one by one or simultaneously due to inter-diffusion, being replaced by the phases enriched more and more in element of higher melting point.

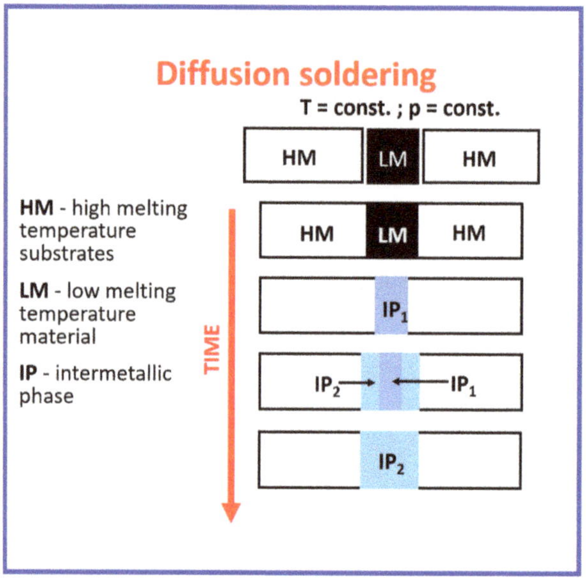

Figure 1. Schematic of diffusion soldering process [22].

In the Ni/Al/Ni system the sequence of appearance of the intermetallics can be predicted based on the Al-Ni equilibrium phase diagram (Figure 2) [25]. The phases are being created going from the lower melting to higher melting component. What is interesting, there are contradictory literature reports concerning the sequence of the intermetallics formation in Ni/Al system, regarding the order of formation of the high aluminum intermetallic phases such as Al_3Ni and Al_3Ni_2. Two opposite approaches are possible. First one involves situation, when as a primary phase Al_3Ni phase precipitates and then Al_3Ni_2 one is created [13,14,26]. Some modeling results and in-situ experiments indicated on formation of Al_3Ni_2 phase as the first one and then Al_3Ni [27–29]. As the process proceeds, the intermetallic compounds such as: AlNi, Al_3Ni_5 and $AlNi_3$ are formed [13]. In the cited works the chemical composition was verified mainly by EDS technique in SEM [13,14,26–29]. Additionally, in [26,29] works, the XRD measurements were also conducted to confirm the phases' composition present in the interconnection zones, while in [27,28] for the phase sequence occurrence the mathematical model was proposed.

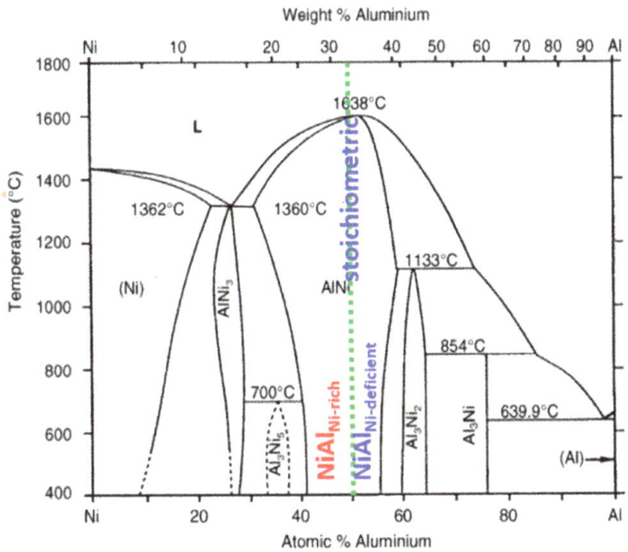

Figure 2. Equilibrium phases diagram Ni–Al with distinction between various types of the AlNi phase [25].

Literature's survey also presents the validity of diffusion studies on Ni substrates of variable grain boundaries. Works of Yu-Chen Tseng et al. [30] based on liquid–solid Sn/Ni system indicated significant differences in the course of the diffusion process conducted with applied commercial coarse-grained Ni substrates and fine-grained Ni layer electroplated (e-Ni) on Ba_2Te_3. In both type of samples after the annealing, the Ni_3Sn_4 phase is formed, but for e-Ni substrate, additional Ni_3Sn_2 phase in the form of the thin film between Ni and Ni_3Sn_4 appeared. A large number of grain boundaries in e-Ni layer, being fast diffusion channels for Ni atoms towards Ni_3Sn_4, resulted in the growth of this additional phase, consisting of more nickel. The additional layer suppressed the growth of Ni_3Sn_4 on e-Ni layer, which is thinner in comparison to the one, formed using coarse grained Ni substrates. A similar experiment in solid Ni/liquid Al/solid Ni system is more difficult to analyze, as higher joining temperature is necessary to apply than the one for Sn/Ni diffusion pair and the nickel recovery and recrystallization phenomena occurs.

This paper shows the results concerning the relation between the reactivity (intermetallic phases growth kinetics) in Ni/Al system and the microstructure of nickel substrates. As the grain boundaries can act as fast diffusion paths, they can influence the growth kinetics of the intermetallic phases and therefore, different results can be reported for apparently the same experimental procedure. In this work Ni substrates of two crystallographic orientations was employed to study the growth of the intermetallic phases in solid/liquid and solid/solid state in Ni/Al/Ni system.

2. Materials and Methods

Substrates used in the experiment were prepared from high purity commercial Ni rod (99.999, (Goodfellow Cambridge Ltd., Huntingdon, UK) with a diameter of 5 mm. The rod was cut in two orientations: along and perpendicular to the rod elongation direction as it is showed in Figure 3. EBSD maps revealed the different crystallographic orientation of both types of substrates. Optical microstructures and EBSD maps show significant difference in the appearance of both substrates, for substrate NiA-type (Figure 3a) grains are elongated and narrow. On the other hand, for the NiB-type substrate (Figure 3b), grain possess irregular shape and on the map considerable refinement of the structure is visible.

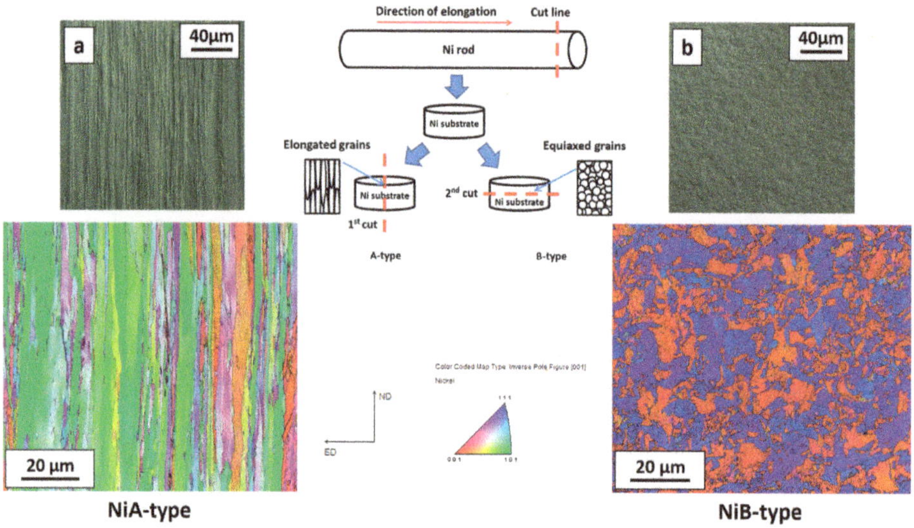

Figure 3. The scheme of the preparation of nickel substrates with their optical microstructures and electron backscattered diffraction (EBSD) maps (**a**) NiA-type, (**b**) NiB-type.

In order to obtain the diffusion-soldered interconnections, two nickel slices with the same or different orientation were grinded, polished and cleaned in ultrasonic cleaner for 300 s. Then the thin (80 µm) slice of high purity aluminum (99.999, Goodfellow Cambridge Ltd., Huntingdon, UK) was clamped between two Ni substrates and held in specific temperature for different periods of time. As the aluminum melting point is of 660 °C, the 720 °C was applied as the joining temperature. Such temperature, higher than the one necessary to melt Al, was chosen for several reasons. First of all, it ensures that Al passes to the liquid state during the annealing in the vacuum. It also allows comparing obtained results with the data presented by Lopez et al. [13]. Only slight mechanical pressure was used to avoid of leakage of the solder and the samples were sealed in quartz ampules to prevent the samples' oxidation. Table 1 shows variety of applied experimental assemblies and conditions of their annealing.

Table 1. The experimental assemblies and conditions. NiA denotes the substrate cut to the parallel direction to the elongation rod, while the NiB in the perpendicular direction.

	Conditions:	**Temperature 720 °C, Vacuum, Slight Mechanical Pressure, Heating 18 °C/min Cooling with Furnace**
No.	**Sample Type**	**Time of Annealing [h]**
1.	NiA/Al/NiB	0.25
2.	NiA/Al/NiB	0.50
3.	NiA/Al/NiA NiB/Al/NiB	1
4.	NiA/Al/NiB	2
5.	NiA/Al/NiA NiB/Al/NiB	3
6.	NiA/Al/NiB	4
7.	NiA/Al/NiA NiB/Al/NiB	5
8.	NiA/Al/NiB	20
9.	NiA/Al/NiB	72

The cross-sections of the interconnections for scanning electron microscopy (SEM, Quanta 3D FEG, FEI, Hillsboro, OR, USA) examinations were prepared by standard metallographic procedure: embedding samples in epoxy, grinding and then polishing with the diamond paste (3 μm) and silica (0.04 μm). As a starting point the scanning electron microscopy observations with 20 kV accelerating voltage and energy dispersive X-ray spectroscopy (EDS, Trident (EDS-EBSD-WDS), EDAX Inc., Tilburg, The Netherlands) analysis were carried out for each sample, revealing its phase composition and thickness of particular intermetallics layers and also the chemical composition changes across them. The samples in SEM were inspected using the backscattered electrons mode (BSE). As a next step, the electron backscattered diffraction technique was used to expose the amount and character of the grain boundaries and to correlate it with the creating phases composition and thickness. The thin foils for the transmission electron microscopy (TEM, TECNAI G2 200 kV, FEI, Hillsboro, OR, USA) observations were prepared using the Focused Ion Beam (FIB, Quanta 2D, FEI, Hillsboro, OR, USA) technique. It is the only technique suitable for this type of thin foils (exact location of the place of interest). During the milling process, problems with uneven consumption of the sample material were encountered. Difficulties were associated with significant difference in hardness of the various intermetallics phases located at the reaction zone. The obtained thin foils with the thickness of about 100 nm were next examined by TEM.

3. Results and Discussion

3.1. Sequence of Intermetallic Phases in Interconnections

The diffusion-soldering at the temperature of 720 °C for different periods of time resulted in the growth of the several intermetallic phases in the joined area. It is important that the sequence of phase-creation in all cases was the same as it was predicted in [13,14,26]. They grew according to equilibrium phases diagram from the ones rich in low melting component to the ones with higher amount of nickel. Sequence of their appearance in the interconnection zone depended on the duration of reaction.

The initial stage of reaction in Ni/Al/Ni interconnection was observed after 15 min of annealing (Figure 4a). SEM observations using BSE mode showed the contrast differences at the Ni/solder interface, pointing the existence of two intermetallic phases. The measurements of the chemical composition within the interconnection zone confirmed that these phases were Al_3Ni (76.0 at. % Al, 24.0 at. % of Ni) and Al_3Ni_2 (60.6 at. % of Al, 39.4 at. % of Ni). On the other hand, the middle of the joined zone was composed of Al_3Ni-Al eutectics (97.1 at. % of Al, 2.9 at. % of Ni) instead of pure aluminum. Moreover, inside of the Al_3Ni-Al eutectics, the primary precipitates of the Al_3Ni intermetallic phase possessing the faced walls could be observed. Thanks to the channeling contrast, the dual-morphology of Al_3Ni_2 is visible, showing the larger grains to be located closer to the middle of the interconnection and finer grains being located close to the nickel substrates. As it is showed in Figure 4a, the Al_3Ni phase, growing next to the nickel substrate, formed the areas of irregular shape at the interface with eutectics. Such a morphology is called scallops and it is typical for the growth of the intermetallics with assistance of the liquid. Additionally, it was noticed that the interface between Al_3Ni and Al_3Ni_2 phases is wavy. At this point it can be summarized that the interconnection consisted of the following constituents:

$$\mathbf{Ni/Al_3Ni_2/Al_3Ni/[(Al)+Al_3Ni]_{eutectics}/Al_3Ni/Al_3Ni_2/Ni.}$$

The interface between nickel and Al_3Ni_2 phase is smooth and approximately parallel to the surface of the applied substrates. This type of planar interfaces is characteristic for solid/solid reaction. Similar observation was noticed in other papers [14,26].

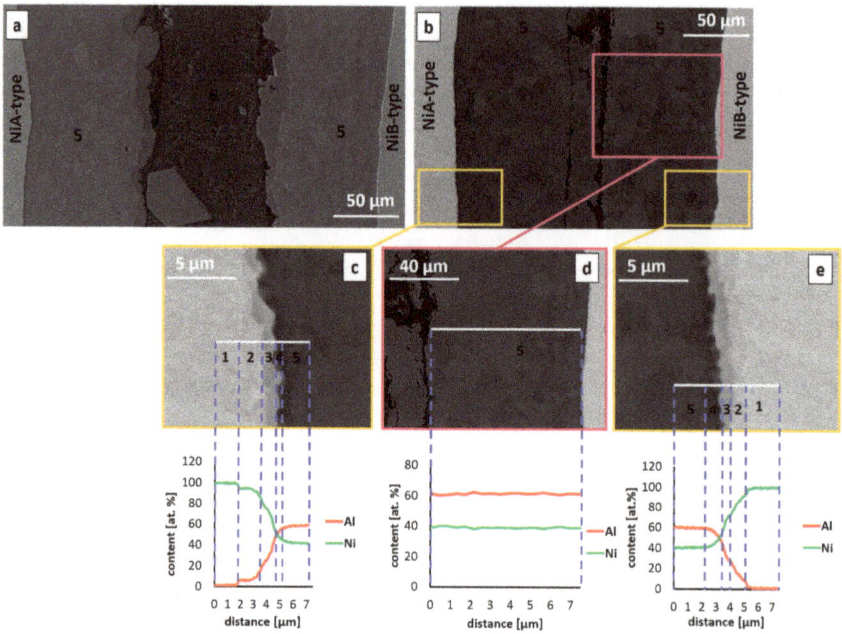

Figure 4. The SEM microstructures after (**a**) 15 min and (**b**) 30 min of reaction at 720 °C. The SEM microstructures in (**c**–**e**) present the magnified area of Al_3Ni_2/Ni interface. Numbers 1–6 denote various phases: 1-Ni, 2-Ni solid solution, 3-$AlNi_3$, 4-AlNi, 5-Al_3Ni_2, 6-(Al)+Al_3Ni.

The obtained interconnection zone after 15 min of annealing at 720 °C may be compared with joint described by Tumminello and Sommadossi [14], for the same time of annealing (15 min) but at higher temperature of 776 °C. The sequence of the intermetallics creation is in both cases compatible. In present work, the interconnection zones are much broader in comparison to the ones shown in [14], being approximately twice as wide. Table 2 contains the comparison of the results from both papers. In present work the whole joint consists of Al_3Ni in 21% and Al_3Ni_2 in 59%, the rest is the eutectics (Al) + Al_3Ni. In Tumminello's work [14] the amount of the phases were 13% and 47%, respectively. The observed differences between these two works could arise mostly from the various thickness of applied aluminum foil (much broader in [14]) and also possibly from the used atmosphere (here vacuum, while argon in [14]).

Table 2. Average thickness of the whole joint and particular intermetallic phases—both sides of the joint.

Phase	Present Work	Tumminello et al. [14]
	Thickness [µm]	
Whole interconnection zone	280	140
Eutectics Al + Al_3Ni	86	56
Al_3Ni	15	9
Al_3Ni_2	82	33

Elongation of the reaction time to 30 min, allowed observing significant difference of the interconnection zones in comparison to the sample after 15 min of annealing. Figure 4b shows the interconnection at this stage of annealing, where the middle of the joint is completely filled by Al_3Ni_2 phase, which was confirmed by EDS. The line scan through the area inside the pink frame, allowed excluding the presence of the eutectics (Al_3Ni–Al) and Al_3Ni phase in this area. The next important microstructural feature exists at the interface between Al_3Ni_2 phase and Ni substrate and has occurred in early stages of solid/solid reaction. The EDS measurements indicate high variability of chemical composition in the areas marked with yellow frames. Average content of aluminum in Al_3Ni_2 equals 60.5 at. %, while in the close neighborhood of the Al_3Ni_2/Ni interface it is about 57.6 at % and decreases in the direction toward the nickel substrate. The next measurements of the chemical composition showed the presence of narrow zones of about 1.5 μm, where the content of aluminum equals 51.3 at. %, which can be attributed to AlNi phase, then 30.3 at. % of Al, standing for $AlNi_3$ phase. The presence of Ni solid solution was also observed (0.8 at. % of Al). In the mentioned area the differences in BSE contrast can be observed, the last one being distinguishable at the Al_3Ni_2/AlNi interface. Changes of the chemical composition near the phases' boundaries indicate progressive diffusion processes leading to the initial stages of the AlNi growth after 30 min of annealing. Interconnection zone constituents can be this time summarized as follows:

$$\mathbf{Ni/Ni_{solidsolution}/AlNi_3/AlNi/Al_3Ni_2/AlNi/AlNi_3/Ni_{solidsolution}/Ni.}$$

The interfaces of the phases growing due to the reaction in the solid state show a linear character, as it was the case in the samples being annealed for 15 min shown in Figure 4a.

Figure 5 shows the morphology of the growing phases after 1, 3 and 5 h of reaction at 720 °C. SEM observations revealed the existence of several intermetallic phases. The measurements of the chemical composition within the joined area indicated that they consisted of:

After 1 h:

$$Ni/Ni_{solidsolution}/AlNi_3/AlNi/AlNi_{Ni\text{-}deficient}/Al_3Ni/[(Al)+Al_3Ni]_{eutectics}/Al_3Ni_2/AlNi_{Ni\text{-}deficient}/AlNi/AlNi_3/Ni_{solidsolution}/Ni$$

After 3 h:

$$Ni/Ni_{solidsolution}/AlNi_3/AlNi_{Ni\text{-}rich}/AlNi/AlNi_{Ni\text{-}deficient}[(Al)+Al_3Ni]_{eutectics}/AlNi_{Ni\text{-}deficient}/AlNi/AlNi_{Ni\text{-}rich}/AlNi_3/Ni_{solidsolution}/Ni$$

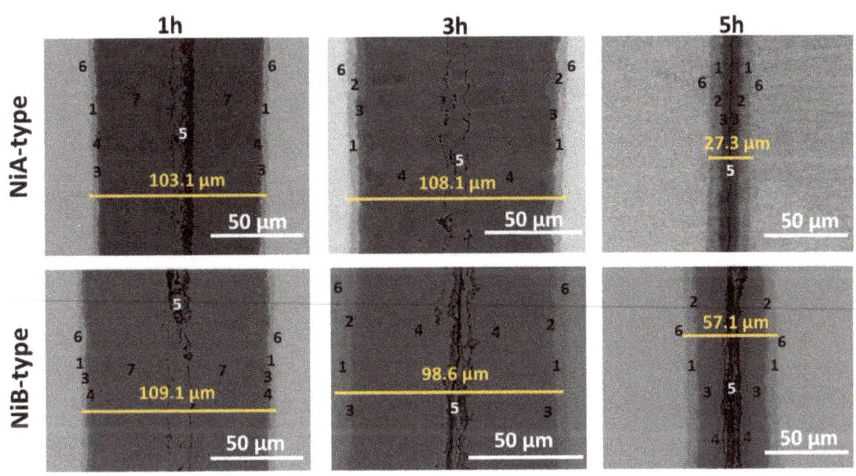

Figure 5. Microstructure of the Ni/Al/Ni couples obtained at 720 °C after annealing for: 1, 3, 5 h using A- and B- type of Ni substrates. Numbers 1–6 denote particular intermetallic phases: 1-$AlNi_3$, 2-$AlNi_{Ni\text{-}rich}$, 3-AlNi, 4-$AlNi_{Ni\text{-}deficient}$, 5-(Al) + Al_3Ni, 6-Ni solid solution, 7-Al_3Ni_2.

Obtained results are the same for both types of applied orientations of nickel substrates. However, after 5 h of reaction time one significant difference was noticed. Namely, the sequence of the intermetallic phases for types A and B was not the same (Figure 6). For Ni/Al/Ni interconnection, where the substrates of type B were used, the phase sequence was the same as in the case of 3 h of annealing, whereas, for the joints obtained from substrates of A-type, the AlNi phase deficient in nickel ($AlNi_{Ni\text{-}deficient}$) did not appear. Figure 5 presents the phase composition of the interconnection zones for both types of substrates, where particular phases (as previously) are noted by numbers. As can be seen in the equilibrium phase diagram presented in Figure 2, the AlNi phase possesses a wide range of chemical composition. Therefore, there are several types of AlNi intermetallics, namely: stoichiometric, where the ratio of Ni to Al equals (50:50 at. %), AlNi deficient in nickel (45–50 at. % of Ni) and AlNi rich in nickel (50–60 at. % of Ni). Last one was evidenced in the literature data by Lopez et al. [13].

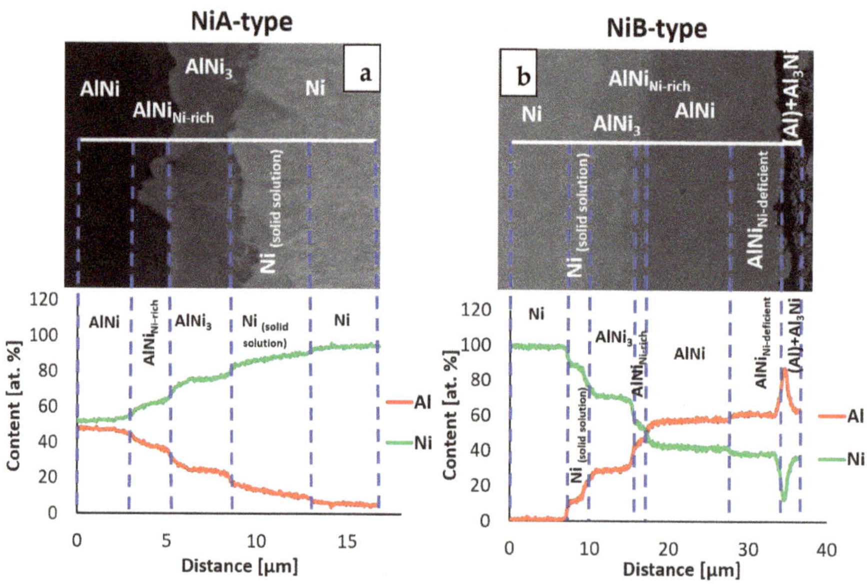

Figure 6. SEM micrographs of the cross-sectional view of a Ni/Al/Ni interconnections after 5 h of reaction time at 720 °C and EDS line-scan across this area for substrates of A-type (**a**) and B-type (**b**).

A significant expansion of the annealing time to 20 and 72 h resulted in widening (Figure 7a) and disappearance (Figure 7b) of particular phases. After 20 h annealing at 720 °C the phases are broadened, however, the location of phases within the interconnection is the same as for samples annealed for 5 h using the substrates of B-type. The interconnection zone is symmetric. The AlNi deficient in nickel is slowly consumed, while the stoichiometric phase expands. 72 h is enough time to fully consume AlNi deficient in nickel and in the interconnection area only AlNi (51.6 at. % Ni), AlNi rich in nickel (60 at. % Ni) and Al_3Ni (24.1 at. % Ni) are present. Phases of AlNi type are approximately twice wider in comparison to 20 h of annealing, $AlNi_3$ phase changes its thickness of about 30% (broadening). The sequence of the phases after 72h is as follow:

$$\mathbf{Ni/Ni_{solidsolution}/AlNi_3/AlNi_{Ni\text{-}rich}/AlNi/AlNi_{Ni\text{-}rich}/AlNi_3/Ni_{solidsolution}/Ni.}$$

Average chemical composition of all intermetallic phases after annealing for different periods of time is collected in Table 3.

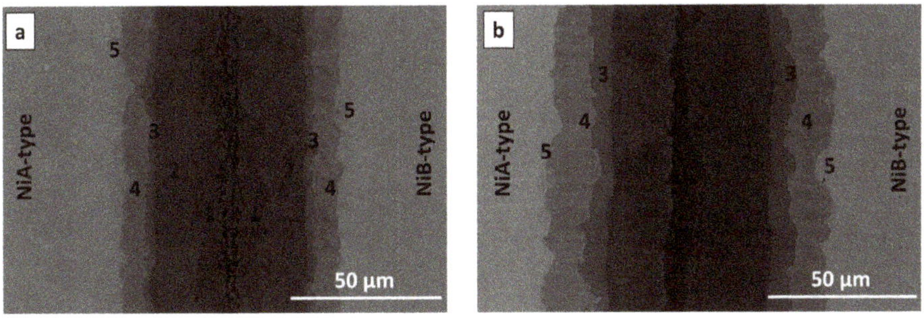

Figure 7. SEM microstructures of the Ni/Al/Ni couples obtained at 720 °C after annealing for: (**a**) 20 and (**b**) 72 h using A- and B- type of substrates of Ni. Numbers 1–5 denote particular intermetallic phases: 1-$AlNi_{Ni\text{-}deficient}$, 2-AlNi, 3-$AlNi_{Ni\text{-}rich}$, 4-$AlNi_3$, 5-Ni solid solution.

Table 3. Chemical composition of the intermetallic phases in Ni/Al/Ni joints after annealing for different periods of time.

Phase	The Range of Aluminum Content (at. %) in Interconnection	Average Content of Aluminum in the Interconnection (at. %)
Ni solid	5.5–12.1 ± 0.2–0.5	8.9 ± 0.2
$AlNi_3$	21.6–30.5 ± 0.4–0.6	26.1 ± 0.5
$AlNi_{Ni\text{-}rich}$	33.4–42.0 ± 0.7–0.8	39.1 ± 0.8
AlNi	47.1–53.5 ± 0.9–1.1	50.6 ± 1.0
$AlNi_{Ni\text{-}deficient}$	53.7–57.8 ± 1.1–1.2	55.9 ± 1.1
Al_3Ni_2	58.2–60.7 ± 1.2	59.6 ± 1.2
Al_3Ni	75.5–76.3 ± 1.5	76.0 ± 1.5

The interfaces, where the solid/solid diffusion occurs were also examined by EBSD technique. Due to sufficient width of particular phases, which grow with time, the sample annealed for 20 h at 720 °C was selected. The EBSD map in Figure 8a indicates the existence of three main areas, which are indexed starting from the nickel as: Ni, $AlNi_3$ and AlNi. The map shows that the zone of interest consists of grains with a random crystallographic orientation and the one large grain of nickel. The differences in size of grains for particular areas are visible. Much finer grains in comparison to other areas are observed for the intermetallic phase identified as $AlNi_3$. On the other hand, in the case of AlNi phases large grains are observed. In both cases range of these phases grain sizes is variable. Complementary to the EBSD map, for the same area the EDS maps were collected (Figure 8b). As can be noticed, these EDS maps show more individual phases compared to EBSD measurement. Area between Al_3Ni and AlNi intermetallic phases is rich in nickel, in comparison to AlNi phase, while at the second side of AlNi phase, the area deficient in nickel is present. This dependency is compatible with SEM micrograph registered in BSE mode for the same area and it is shown in Figure 9. For full understanding of the observed relation, the EBSD indexing confidence map was imposed in the EDS maps for Al and Ni elements (Figure 8c). The result was surprising, as it was mentioned above, the EBSD map does not indicate the existence of more than three phases. However, the combination of EBSD and EDS maps reveals the concentration gradient throughout the AlNi phase grains. Some grains of AlNi phase are enriched in nickel, confirming previous suspicion of existence of AlNi rich phase. The most interesting observation is that the AlNi Ni-rich phase does not create new grains but rather changes the composition of the grains of already existing phases. In contrary to this, AlNi deficient in nickel possess own grains, being separated from the AlNi stoichiometric phase ones.

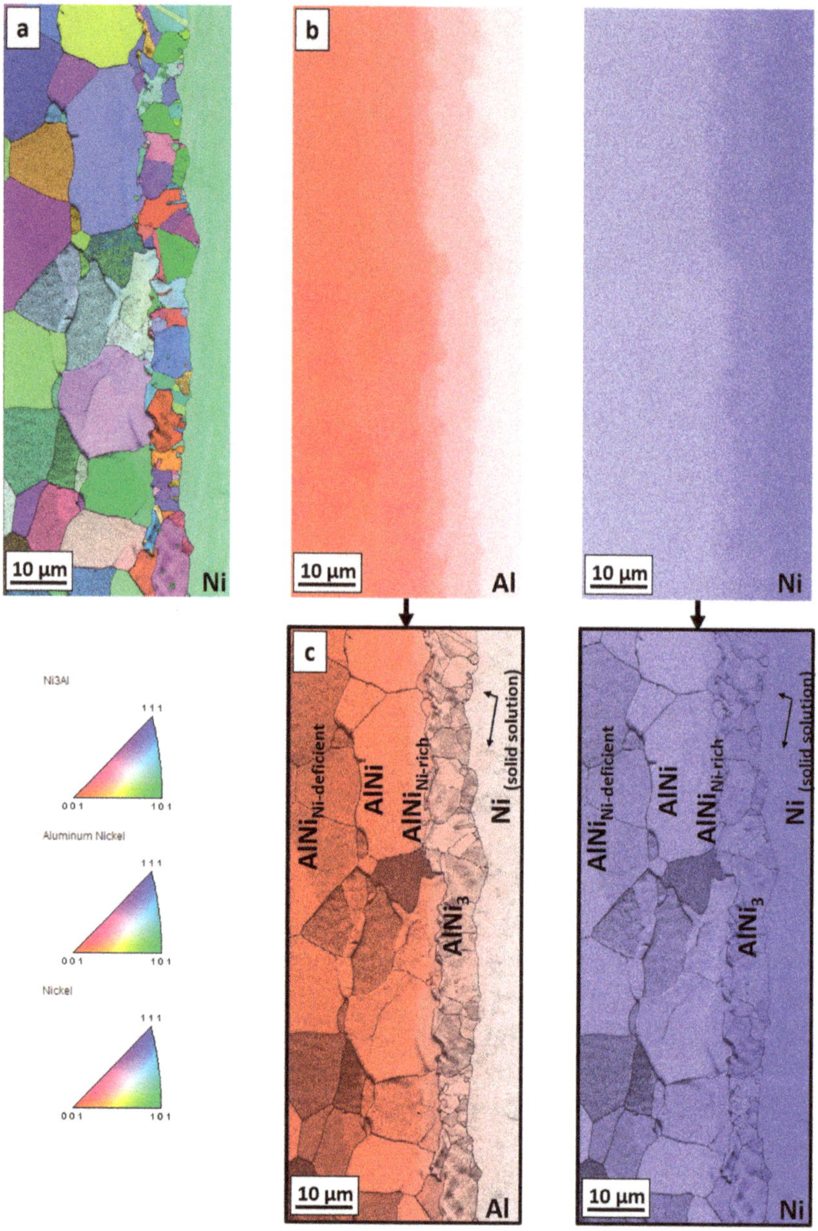

Figure 8. (**a**) EBSD map with (**b**) EDS maps of Al and Ni elements distribution for the sample annealed for 20 h at 720 °C. (**c**) The imposition of the EBSD indexing confidence map and EDS map analysis for Al and Ni elements.

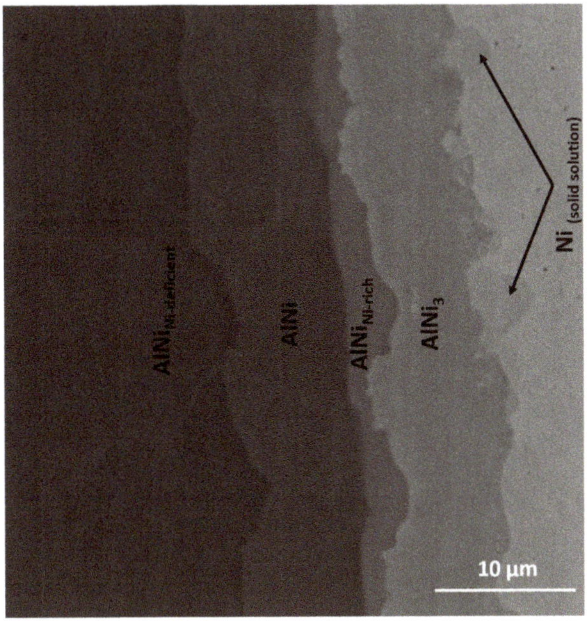

Figure 9. SEM micrograph in BSE mode of the intermetallic phases formed in the solid state after 20 h at 720 °C.

The TEM investigation for sample annealed for 3 h at 720 °C indicates different results in comparison to the EBSD and EDS overlapping for sample after 20 h of annealing. The diffraction pattern from the area taken by the $AlNi_{Ni\text{-}rich}$ phase (Figure 10), determined based on the chemical composition, did not unambigously confirmed its presence. Two possible phases were taken into consideration, namely AlNi and Al_3Ni_5. Important is that AlNi is always identified based on the same crystallographic data [31], however, the content of elements is different, so this phase is considered as AlNi without division into rich and deficient in nickel types. The degree of mismatch is very high for AlNi phase reaching 38% (Figure 10c). For the second considered phase Al_3Ni_5, the degree of mismatch is of only 9% (Figure 10d). This orthorombic phase is metastable below 700 °C, as the samples were cooled with furnace after the annealing process, it could be formed. However, due to the fact that growth of this phase takes place at 720 °C, in the manuscript it is noted as Ni-rich AlNi phase. Further TEM investigations would be of great benefit for the description and understanding of the phase evolution, especially for the early stage of the Ni-rich AlNi phase growth. As the phase thickness was below the analytical resolution in SEM (see Figure 11) such examination in TEM is of essential need. The results of TEM-EDS and SEM-EDS measurements are similar and were collected in Table 4, however, in case of the doubtful phase substantial difference is visible. From SEM-EDS it follows that the phase present in the joint is the Al_3Ni_5 but the TEM-EDS results point at the AlNi rich in nickel one. As mentioned above, this phase was extremely narrow, the thickness is on the border of the resolving power of the method. Examined areas for both methods are shown in Figure 11.

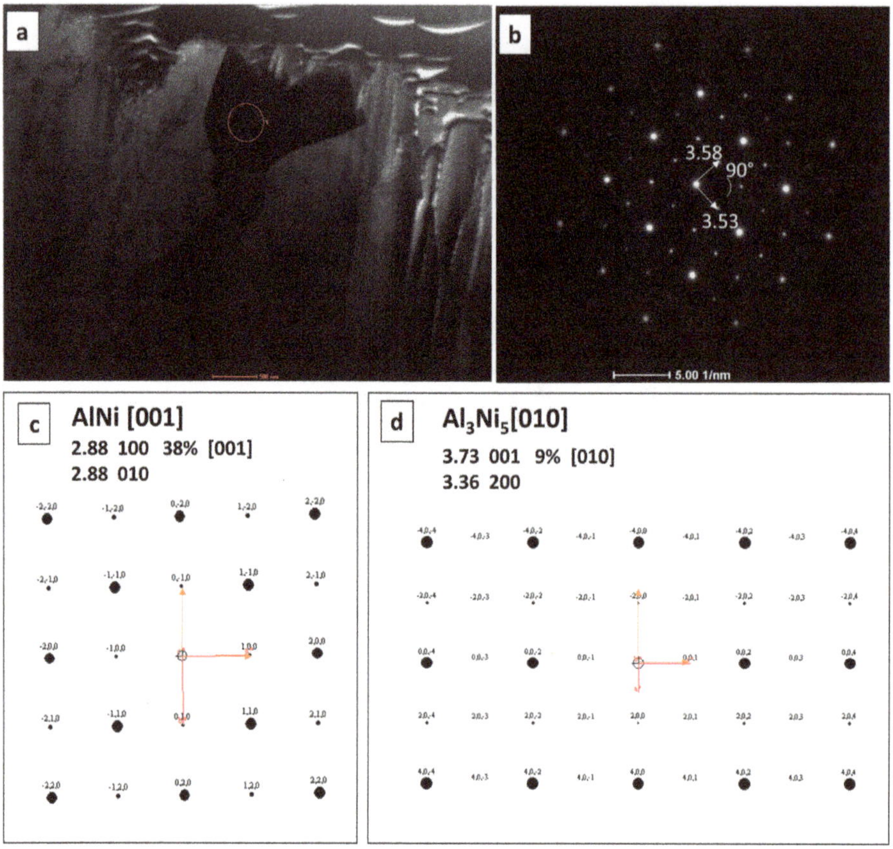

Figure 10. (**a**) TEM bright field image showing the microstructure of the AlNi intermetallic phase for sample annealled for 3 h at 720 °C together with the corresponding (**b**) selected area diffraction pattern taken from the grain marked with circle. Simulation of the solve for (**c**) AlNi and (**d**) Al_3Ni_5.

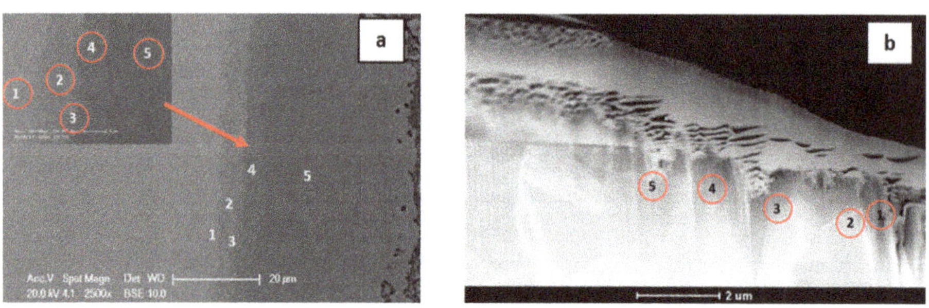

Figure 11. SEM (**a**) and TEM (**b**) microstructures of a Ni/Al/Ni interconnections (Ni substrate of B-type) after 3 h of reaction time at 720 °C with indicated EDS point analysis presented in Table 4. Numbers 1–5 denote particular intermetallic phases: 1-Ni solid solution, 2-$AlNi_3$, 3-$AlNi_{Ni\text{-}rich}$, 4-AlNi, 5-$AlNi_{Ni\text{-}deficient}$.

Table 4. Comparison of the chemical compositions obtained by SEM and TEM.

No.	Method	SEM		TEM	
	Phase	Content at. %			
		Ni	Al	Ni	Al
1.	Ni solid solution	94.4 ± 1.9	5.6 ± 1.9	91.9 ± 1.8	8.1 ± 1.8
2.	$AlNi_3$	78.4 ± 1.6	21.6 ± 1.6	78.4 ± 1.6	21.6 ± 1.6
3.	$AlNi_{Ni\text{-}rich}$	66.6 ± 1.3	33.4 ± 1.3	60.8 ± 1.2	39.2 ± 1.2
4.	AlNi	50.1 ± 1.0	49.9 ± 1.0	51.9 ± 1.0	48.1 ± 1.0
5.	$AlNi_{Ni\text{-}deficient}$	45.6 ± 1.0	54.4 ± 1.0	45.8 ± 0.9	54.2 ± 0.9

3.2. Growth Rate of the Intermetallic Phases

Determination of the thickness of the intermetallics (Table 5) allowed revealing their growth kinetics. For the short time of the reaction, the interconnection zones were broadening as it is assumed in the diffusion soldering process. Between 1 and 3 h of annealing only the subtle difference of the thickness of the whole joint was observed, nevertheless, different intermetallic phases in reaction zones appeared. Changes of phases composition points that the reaction after 1 h probably takes place in the solid state. After 5 h of annealing, due to isothermal solidification stage of DS, for both types of substrates, the interconnection zones shrink. As it was mentioned earlier, phases composition between 3 and 5 h stays the same, while the main difference is associated with the thickness of the interconnection zones for both types of substrates. For samples of A-type, the interconnection zones after 5 h of annealing is four times narrower and for B-type two times thinner than after 3 h of annealing. Comparing the samples after the same annealing conditions, for A-type to B-type Ni substrates, the following results are observed: after 1 h of annealing the thickness of the entire joint is comparable and widths of particular phases are similar. When time of annealing is extended to 3 h, the differences in diffusion process and in the overall appearance of the interconnection zones are not observed. After 5 h of annealing more visible differences appeared. First of all, the thickness of the joint, where substrate B-type was used, is twice broader than in Ni/Al/Ni reaction zone with A-type substrates. Additionally, in case of NiA/Al/NiA, the phase AlNi deficient in nickel is not observed. Thicknesses of the individual phases are similar, beside of the total width of AlNi phases (of every type), which in case of B-type nickel is broader, however, stoichiometric type of AlNi phases are comparable. This difference between thickness of whole joint after 5 h could be caused by the leakage of liquid solder during the experiment due to too high pressure applied. The comparison of the thickness of the particular layers of the intermetallic phases formed in Ni/Al/Ni interconnection in different time of reaction is collected in Table 5. Authors conducted experiment which allowed to eliminate the necessity of application of two separated systems: NiA/Al/NiA and NiB/Al/NiB. Numerous attempts prove that the localization of the Ni substrates (above or below the Al solder) does not affect the width and sequence of created phases. This approach resulted in simplifying the experimental procedure and allowed for producing of NiA/Al/NiB system (and conversely), which shortened experiment time. This procedure was used for shorter and longer annealing times. In early stages of diffusion soldering processes phase Al_3Ni disappears fast—only after 30 min it is completely consumed and replaced by Al_3Ni_2 phase, which, in turn, after 3 h of annealing no longer exists. The longest times of annealing cause further phases broadening. Finally, after 72 h of annealing the interconnection zone contains only high nickel phases (50 at.% of Ni and higher). Three phases: $AlNi_3$, AlNi, $AlNi_{Ni\text{-}rich}$ for which the growth kinetic is calculated expand gradually with the annealing time.

Table 5. The thickness of the particular layers of the intermetallic phases formed in Ni/Al/Ni interconnection in different time of reaction at 720 °C.

Time [h]	Ni-Type	Layer Thickness, Δd [μm]					
		Al_3Ni	Al_3Ni_2	$AlNi_{Ni\text{-}deficient}$	AlNi	$AlNi_{Nirich}$	$AlNi_3$
0.25	A	16.9	87.8	-	-	-	-
	B	12.9	76.5	-	-	-	-
0.50	A	-	94.2	-	0.7	-	0.7
	B	-	96.3	-	0.8	-	0.8
1	A	-	40.2	-	1.9	-	2.5
	B	-	50.4	-	2.2	-	2.5
2	A	-	108.7	-	2.5	1.0	2.7
	B	-	103.7	-	1.7	1.0	2.8
3	A	-	-	48.1	2.9	1.1	3.3
	B	-	-	38.1	4	1.3	4.1
4	A	-	-	77.6	3	1.1	3.4
	B	-	-	45.5	2.7	1.5	4.2
5	A	-	-	-	6.5	1.5	5.7
	B	-	-	7.2	10.6	2.5	6
20	A	-	-	7.5	14.2	2.6	9.3
	B	-	-	10.5	13.1	3.3	9
72	A	-	-	-	19.9	6.0	11.2
	B	-	-	-	20.3	7.1	12.7

Determination of the main mechanisms, which control the growth of the intermetallic phase is based on simple and useful formula (Equation 1), in details discussed in [32], giving the relation between the thickness of the intermetallics with the time of annealing.

$$\Delta d = kt^n \quad (1)$$

where: Δd is a thickness of the intermetallic phase layer, k—the growth rate constant and t—time of annealing. To define the mechanism of the intermetallic phase growth, it is necessary to determine the value of n exponent. Depending on this value, the growth can be controlled either by the volume diffusion ($n = 0.5$), or by the chemical reaction at the interfaces ($n = 1$), grain boundary diffusion ($n < 0.5$) and finally by mixed mechanism of growth ($0.5 < n < 1$). The growth rate constant k can be determined from two types of plot, namely Δd vs. $t^{(\frac{1}{2})}$ or Δd^2 vs. t. First type of plot is better for the growth of the intermetallic phases, where the layer is formed at the initial period of growth and does not influence the course of further stabilized growth. Second type should be used, when the initial growth of the obtained layer affects the period of the parabolic growth [32]. Diffusion process may be controlled by the dislocation mechanism but only at lower temperature, while at the temperature close to the melting point of metals the amplitude of thermal vibrations of atoms is too high and dislocations as a structural effects disappear. Therefore, in such a case, the dislocations are not taken into account and the mechanisms of diffusion are either volume diffusion or diffusion by grain boundaries or reactive diffusion. The growth kinetics results are collected in Table 6 and shown in Figure 12.

Table 6. The growth kinetics of AlNi, $AlNi_{Ni\text{-}rich}$ and $AlNi_3$.

Ni A-Type	n	k	Mechanism	Ni B-Type	n	k	Mechanism
AlNi	0.67	$1.51 \frac{\mu m^{0.67}}{h}$	mixed: volume diffusion and chemical reaction at interface	AlNi	0.65	$1.65 \frac{\mu m^{0.65}}{h}$	mixed: volume diffusion and chemical reaction at interface
$AlNi_{Ni\text{-}rich}$	0.51	$1.59 \frac{\mu m^{0.51}}{h}$	volume diffusion	$AlNi_{Ni\text{-}rich}$	0.52	$1.28 \frac{\mu m^{0.52}}{h}$	volume diffusion
$AlNi_3$	0.51	$1.77 \frac{\mu m^{0.51}}{h}$	volume diffusion	$AlNi_3$	0.50	$1.96 \frac{\mu m^{0.50}}{h}$	volume diffusion

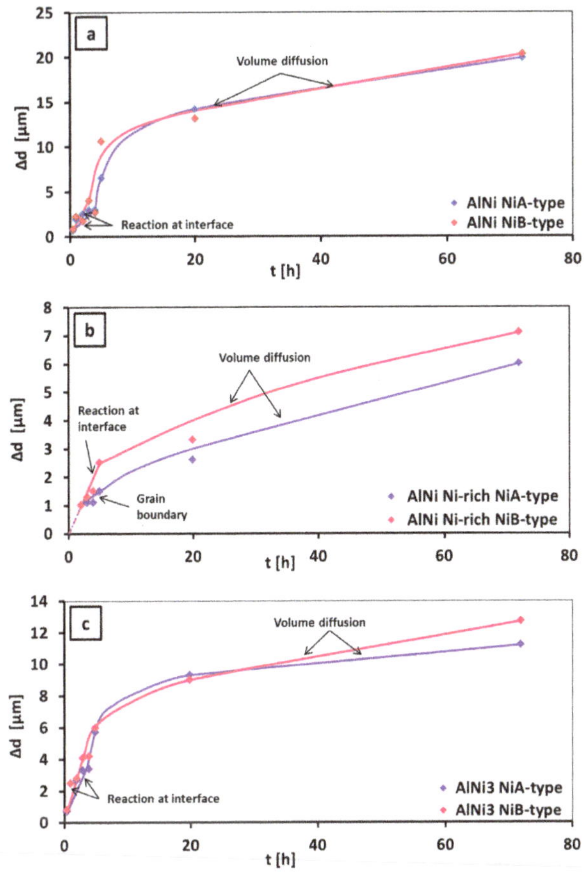

Figure 12. Time dependence of the layer width for stoichiometric AlNi (**a**), $AlNi_{Ni\text{-}rich}$ (**b**) and $AlNi_3$ (**c**) in Ni/Al/Ni joints.

The growth kinetics data for three phases formed in the solid state in Ni/Al/Ni interconnections was determined. Plots (Figure 12) Δd vs. t for AlNi stoichiometric (Figure 12a), $AlNi_{Ni\text{-}rich}$ (Figure 12b) and $AlNi_3$ phases (Figure 12c), showed that the growth of AlNi phases was controlled by different mechanism in comparison to $AlNi_{Ni\text{-}rich}$ and $AlNi_3$. The growth of AlNi phase involved two mechanisms: at first, the reaction at the interface took place and then it was replaced by the volume diffusion. The time exponent n for AlNi phase equals 0.67 for substrates of A-type and 0.65 for the B-type ones. The calculation showed that growth mechanism for AlNi rich in nickel and $AlNi_3$ are similar and governed by the volume diffusion (n is 0.5 ± 0.1). In the case of AlNi rich in nickel phase growth with the short incubation time occurred only for substrate of A-type. The $AlNi_3$ phase grew due to the volume diffusion mechanism and no incubation time was observed in its growth. Finally, it was verified, that neither of these phases grew due to the grain boundary diffusion, therefore, the differences between the samples of A and B Ni substrates were not observed. In the study made by Lopez et al. [13] phases grew only by volume diffusion, while in present work the growth mechanism for the AlNi stoichiometric is found to be mixed (chemical reaction and volume diffusion). Lopez et al. [13] determined that the fastest growth occurred for stoichiometric AlNi and the slowest growth for Ni-rich AlNi one. They also calculated that the values of n factor equaled 0.5 ± 0.1 for all three phases: AlNi, $AlNi_{Ni\text{-}rich}$ and $AlNi_3$. Therefore, the authors assumed that the growth of all these layers obeys a parabolic law—it is governed by the volume diffusion. What is interesting, they also noticed a transition period of $AlNi_3$ growth for short time of annealing. In this study presented graphs in Figure 12 and Table 6 revealed new insight for the Al/Ni interaction. At present work the differences in behavior of phase-growth at short time of annealing and after longer time was noticed. Growth of AlNi rich in Ni phase is governed by volume diffusion but only after longer time of annealing. Focusing on the shorter time of process (2–5 h), the calculated n exponent shows that for substrates of A-type, the grain boundary mechanism dominates ($n = 0.37$). On the other hand, considering of B-type substrate, data obtained at the beginning of DS suggesting the significant contribution of the reaction at interface ($n = 0.91$). Similar behavior for $AlNi_3$ phase for both types of substrates was observed, however in the range of 0.5–5 h time of annealing, only reaction at interface was evidenced as the governing mechanism of growth (for A-type $n = 0.73$; for B-type $n = 0.76$). In the case of AlNi stoichiometric, independently on time of annealing, two mechanisms of growth, as it was mentioned above - reaction at interface and volume diffusion occurred.

4. Conclusions

The diffusion soldering process was successfully applied to join nickel substrates using aluminum as the solder material at 720 °C for various periods of time. Nickel substrates, applied in the experiment, differed by size and shape of their grains in order to reveal any influence on the sequence of appearance or growth kinetics of particular intermetallic phases. It was demonstrated that at first the intermetallic phases rich in aluminum were formed (Al_3Ni, Al_3Ni_2) and later replaced by the rich in nickel phases (AlNi, $AlNi_3$). Three variants of AlNi phase were identified, all of them clearly visible in SEM-BSE: deficient in nickel (45 at. % of Ni), stoichiometric (50 at. % of Ni) and rich in nickel (60 at. % of Ni). While the Ni deficient AlNi phase formed separated layer of grains, electron backscattered diffraction map evidenced that the Ni-rich variant was located within the grains of already formed AlNi stoichiometric phase. Moreover, transmission electron microscopy examination revealed local presence of Al_3Ni_5 phase grains. Growth kinetics data showed that the fastest growing phase was the stoichiometric variant of AlNi phase, growing in a mixed mechanism: at first due to the chemical reaction and later by the volume diffusion. The most interesting results were obtained for two other phases: Ni-rich AlNi and $AlNi_3$. As it was already showed in the literature, the general growth can be attributed to the volume diffusion, however, at early stages significant differences were noticed in this study. In the case of $AlNi_3$ phase, the first stage of growth was due to the chemical reaction and later it slowed down (volume diffusion). For the Ni-rich AlNi phase in the later stage of growth the volume diffusion dominated. However, the first stage of phase growth varied in relation to the applied nickel substrate. In the interconnections, for which the surface of contact with aluminum was composed of the elongated grains, the grain boundary diffusion mechanism occurred, while for the Ni with small, equiaxed grains, the chemical reaction governed the growth of the phase. This work demonstrates that different mechanisms may contribute in the growth of the phases, especially in the beginning of the process and one should be aware of this. The substrate microstructure may have the influence on the intermetallics' growth kinetics and resulting from it differences in the integral diffusion coefficients.

Author Contributions: Conceptualization, J.W.-B. and I.K.; methodology, I.K., P.B., A.W.-M., L.L.-D., J.W.-B.; formal analysis, I.K., P.B., J.W.-B.; investigation, I.K., P.B., L.L.-D., J.W.-B.; writing—original draft preparation, I.K.; writing—review and editing, J.W.-B.; visualization, I.K.; supervision, J.W.-B.

Funding: The research was co-financed by the European Union from resources of the European Social Fund (Project No.WND-POWR.03.02.00-00-I043/16).

Acknowledgments: The research was co-financed by the European Union from resources of the European Social Fund (Project No.WND-POWR.03.02.00-00-I043/16).

Conflicts of Interest: The authors declare no conflict of interest.

References

1. Miracle, D.B. The physical and mechanical properties of NiAl. *Acta Metal. Mater.* **1993**, *41*, 649–684. [CrossRef]
2. Liu, C.T.; Stiegler, J.O. Ductile ordered intermetallic alloys. *Science* **1984**, *226*, 636–642. [CrossRef] [PubMed]
3. Jung, S.B.; Minamino, Y.; Yamane, T.; Saji, S. Reaction diffusion and formation of Al_3Ni and Al_3Ni_2 phases in the Al-Ni system. *J. Mater. Sci. Lett.* **1993**, *12*, 1684–1686. [CrossRef]
4. Bradley, A.J.; Taylor, A. An X-ray of the nickel-aluminium system. *Proc. R. Soc. Lond.* **1937**, *159*, 56–72.
5. Kodentsov, A. Interdiffusion studies in β- and γ′-intermetallic phases of the binary Ni-Al system. *Diffus. Fund.* **2017**, *13*, 56–97. [CrossRef]
6. Shankar, S.; Seigle, L.L. Interdiffusion and intrinsic diffusion in the NiAl (δ) phase of the AL-Ni system. *Metal. Trans. A* **1978**, *9A*, 1467–1476. [CrossRef]
7. Yang, T.Y.; Wu, S.K.; Shiue, R.K. Interfacial reaction of infrared brazed NiAl/Al/NiAl and $Ni_3Al/Al/Ni_3Al$ joints. *Intermetallics* **2001**, *9*, 341–347. [CrossRef]
8. Tsao, C.L.; Chen, S.W. Interfacial reaction in the liquid diffusion couples of Mg/Ni, Al/Ni, Al/(Ni)-Al_2O_3 systems. *J. Mater. Sci.* **1995**, *30*, 5215–5222. [CrossRef]
9. Castleman, L.S.; Froot, H.A. Nucleation of Intermetallic Phases in Aluminium-Nickel Diffusion couples. *Trans. ASM* **1963**, *56*, 205–208.
10. Lieblich, M.; Gonzalez-Crrasco, J.L.; Caruna, G. Thermal stability of an Al/Ni_3 composite processed by powder metallurgy. *Intermetallics* **1997**, *5*, 515–524. [CrossRef]
11. Gong, X.; Ma, Y.; Guo, H.; Gong, S. Effect of thermal cycling on microstructure evolution and elements diffusion behavior near the interface of Ni/NiAl diffusion couple. *J. Alloy. Compd.* **2015**, *642*, 117–123. [CrossRef]
12. Fujiwara, K.; Horita, Z. Measurement of intrinsic diffusion coefficients of Al and Ni in Ni_3Al using Ni/NiAl diffusion couples. *Acta Mater.* **2002**, *50*, 1571–1579. [CrossRef]
13. Lopez, G.A.; Sommadossi, S.; Zieba, P.; Gust, W.; Mittemeijer, E.J. Kinetic behavior of diffusion-soldered Ni/Al/Ni interconnections. *Mater. Chem. Phys.* **2002**, *78*, 459–463. [CrossRef]
14. Tumminello, S.; Sommadossi, S. Growth Kinetics of Intermetallic Phases in Transient Liquid Phase Bonding Process (TLPB) in Al/Ni System. *Defect. Diffus. Froum.* **2012**, *323–325*, 465–470. [CrossRef]
15. Urrutia, A.; Tumminello, S.; Lames, D.G.; Sommadossi, S. X-Ray Characterization of intermetallic phases in Al/Ni multilayer system. *Proc. Mat. Sci.* **2015**, *8*, 1150–1159. [CrossRef]
16. Maj, L.; Morgiel, J. In-situ transmission electron microscopy observations of nucleation and growth of intermetallic phases during reaction of Ni(V)/Al multilayers. *Thin Solid Films* **2017**, *621*, 165–170. [CrossRef]
17. Maj, L.; Morgiel, J.; Mars, K.; Grzegorek, J.; Faryna, M.; Godlewska, E. Microstructure and hardness of Ti6Al4V/NiAl/Ti6Al4V joints obtained through resistive heating. *J. Mater. Process. Tech.* **2018**, *255*, 689–695. [CrossRef]
18. Simoes, S.; Viana, F.; Kocak, M.; Ramos, A.S.; Viera, M.T.; Viera, M.F. Diffusion bonding of TiAl using reactive Ni/Al nanolayers and Ti and Ni foils. *Mater. Chem. Phys.* **2011**, *128*, 202–207. [CrossRef]
19. Konieczny, M. Mechanical properties and deformation behavior of laminated Ni-(Ni_2Al_3+$NiAl_3$) and Ni-(Ni_3Al+NiAl) composites. *Mater. Sci. Eng.* **2013**, *586*, 11–18. [CrossRef]
20. Noro, J.; Ramos, A.S.; Viera, M.T. Intermettallic phase formation in nanometric Ni/Al multilayer thin films. *Intermetallics* **2008**, *16*, 1061–1065. [CrossRef]
21. Brunelli, K.; Peruzzo, L.; Dabala, M. The effect of prolonged heat treatments on the microstructuralevolution of Al/Ni intermetallic compounds in multi layered composites. *Mater. Chem. Phys.* **2014**, *149–150*, 350–358.

22. Wojewoda, J.; Lopez, G.A.; Zieba, P.; Mittemeijer, E.J. Diffusion processes in diffusion-soldered interconnections. *Arch. Metall. Mater.* **2004**, *49*, 277–291.
23. Humpston, G.; Jacobson, D.M.; Sangha, S.P.S. Diffusion soldering for electronics manufacturing. *Endeavour* **1994**, *18*, 55–60. [CrossRef]
24. Zhou, Y. Introduction to diffusion soldering/brazing. In *Microjoining and Nanojoining*; Zhou, Y., Ed.; Woodhead Publishing Limited: Cambridge, UK, 2008; Volume 1, pp. 269–298.
25. Nash, P.; Singleton, M.F.; Murray, J.L. *Phase Diagrams of Binary Nickel Alloys*; Nash, P., Ed.; ASM International: Materials Park, OH, USA, 1991; pp. 3–11.
26. Bouche, K.; Barbier, F.; Coulet, A. Phase Formation During Dissolution of Nickel in Liquid Aluminium. *Z. Metallkd.* **1997**, *88*, 446–451.
27. Wolczynski, W.; Guzik, E.; Janczak-Rusch, J.; Kopycinski, D.; Golczewski, J.; Mo Lee, H.; Kloch, J. Morphological characteristics of multi-layer/substrate systems. *Mater. Charact.* **2006**, *56*, 274–280. [CrossRef]
28. Wolczynski, W.; Okane, T.; Senderowski, C.; Kania, B.; Zasada, D.; Janczak-Rusch, J. Meta-Stable Conditions of Diffusion Brazing. *Arch. Metall. Mater.* **2011**, *56*, 311–323. [CrossRef]
29. Ding, Z.; Hu, Q.; Lu, W.; Sun, S.; Xia, M.; Li, J. In situ observation on the formation of intermetallics compounds at the interface of liquid Al/ solid Ni. *Scripta Mater.* **2017**, *130*, 214–218. [CrossRef]
30. Tseng, Y.C.; Lee, H.; Tsai, S.C.; Yen, Y.W.; Chen, C.M. Suppression effect of Ni grain size on the Ni_3Sn_4 growth at the Sn/Ni interface. *Mater. Charact.* **2017**, *128*, 232–273. [CrossRef]
31. Ainutdinov, F.A.; Khairidinov, S.K.; Vakhobov, A.V. *Doklady Akademii Nauk Tadzhiskoj SSR 3*; Akademija Nauk SSR: Moscow, Russia, 1987; Volume 30, pp. 169–172.
32. Pieraggi, B. Calculations of Parabolic Reaction Rate Constants. *Oxid. Met.* **1987**, *27*, 177–185. [CrossRef]

nanomaterials

MDPI

Article

Effects of Residual Stress Distribution on Interfacial Adhesion of Magnetron Sputtered AlN and AlN/Al Nanostructured Coatings on a (100) Silicon Substrate

Rashid Ali [1], Marco Renzelli [2], M. Imran Khan [1], Marco Sebastiani [3,*] and Edoardo Bemporad [3]

1 Faculty of Materials and Chemical Engineering, Ghulam Ishaq Khan Institute of Engineering Sciences and Technology, Topi, Swabi 23640, KPK, Pakistan; rashidali@giki.edu.pk (R.A.); imrankhan@giki.edu.pk (M.I.K.)
2 LFoundry s.r.l., via Pacinotti, 7 Avezzano, 67051 L'Aquila, Italy; renzelli.marco@gmail.com
3 Engineering Department, Università degli studi Roma Tre, via dellaVasca Navale, 79, 00146 Rome, Italy; edoardo.bemporad@uniroma3.it
* Correspondence: seba@uniroma3.it; Tel.: +39-06-57333303

Received: 17 October 2018; Accepted: 29 October 2018; Published: 1 November 2018

Abstract: The present study investigated the influence of nanoscale residual stress depth gradients on the nano-mechanical behavior and adhesion energy of aluminium nitride (AlN) and Al/AlN sputtered thin films on a (100) silicon substrate. By using a focused ion beam (FIB) incremental ring-core method, the residual stress depth gradient was assessed in the films in comparison with standard curvature residual stress measurements. The adhesion energy was then quantified by using a nanoindentation-based model. Results showed that the addition of an aluminum layer gave rise to additional tensile stress at the coating/substrate interface, which can be explained in terms of the differences of thermal expansion coefficients with the silicon substrate. Therefore, the coatings without the Al layer showed better adhesion because of a more homogeneous compressive residual stress in comparison with the coating having the Al layer, even though both groups of coatings were produced under the same bias voltage. Results are discussed, and some general suggestions are made on the correlation between coating/substrate property combinations and the adhesion energy of multilayer stacks. The results suggested that the Al bond layer and inhomogeneous residual stresses negatively affected the adhesion of AlN to a substrate such as silicon.

Keywords: physical vapor deposition; magnetron sputtering; AlN/Al coating; silicon substrate; residual stresses; wafer curvature method; nanoscale residual stress profiling; indentation failure modes; nanoindentation adhesion

1. Introduction

High-quality thin coatings of aluminium nitride (AlN) have been extensively used in electronics for heat dissipation applications [1]. These coatings have been produced by several established methods. However, due to simplicity, reproducibility, and lower cost, magnetron sputtering remains one of the common techniques used for the deposition of AlN coatings. One of the peculiar features of many physical vapor-deposited (PVD) coatings is the presence of high compressive residual stresses. These stresses are an intrinsic outcome and could be a limitation for the maximum achievable thickness and/or for optimized adhesion to the substrate.

In a recent review article [2], the origin and evolution of residual stress in thin films is discussed in detail and the correlations between process parameters, microstructures, and stress distributions are analyzed. Here, it is shown how complex residual stress depth gradients can be generated during the growth of PVD thin films. Those stresses, even if compressive on average, can have an unwanted effect on the mechanical response of thin films during service. This is also true in case of Atomic Layer Deposition (ALD) dielectric layers (which can be denser and perfectly conformal to the substrate, in comparison with PVD), where the presence of residual stress can generate cracks [2]. Recent papers by some of the authors [3,4] showed that the presence of a controlled residual stress gradient in multilayer Cr–CrN PVD thin films can have a very beneficial effect on scratch adhesion and can be a powerful design tool for nanostructured coatings with improved performance. Therefore, one of the most important avenues of research for magnetron sputtered PVD coatings is to increase the adhesion of the coating to the substrate under optimized residual stress gradients [3,4].

Similar to the commonly used Rockwell-C indentation and scratch testing techniques for determining the adhesion of a thick hard coating on a ductile substrate, nanoindentation is also used as an enabling technique to investigate the adhesion of thin coating systems onto hard substrates. Especially in the case of very thin nanostructured layers (<1 μm), standard scratch and HRC tests fail to output reliable adhesion values, since the response of the coating/substrate system becomes a strong function of the substrate's properties and may change remarkably as a function of the adopted indenter tip. There are several additional benefits that make nanoindentation attractive; only a small sample is needed to perform the test that can potentially supply several materials properties, such as hardness, modulus, and adhesion [5]. There are few studies in the authors' knowledge that quantitatively investigate the effect of residual stresses on contact-induced coating adhesion and delamination mechanisms. A quantitative measurement of coating adhesion is essential to allow their exploitation for many critical applications, particularly within the semiconductor microelectronics industry [6,7]. This paper reports the work that aimed to clarify the role of interfacial residual stresses in the adhesion of thin films, as measured by nanoindentation coupled with scanning electron microscopy (SEM) investigations of indentation-induced failure modes.

To this purpose, four different coating systems were produced: a simple AlN single layer on a Si substrate, where the film was expected to be in compression; the same AlN layer with the addition of an aluminum bond layer that was designed to have tensile residual stress in the aluminum layer (as it usually happens for Al metallic sputtered layers), to investigate the effect of an inhomogeneous state of stress (tensile–compressive) on the adhesion of the coating. The same AlN coatings (with and without an Al bond layer) were also deposited under higher bias voltage conditions, in order to investigate the effects of an increased compressive residual stress in the top layer on adhesion. Average residual stress and through-thickness stress gradient were measured for their influence on interfacial adhesion. Through-thickness distribution of the residual stress gradient and mechanical properties are essential for characterizing nanostructured materials [8].

Finally, by combining the results from the four different samples, a discussion was made on how the residual stress distribution in Al/AlN sputtered coatings would affect adhesion and crack propagation modes.

2. Experimental Details

Coatings were produced using a standard PVD magnetron sputtering plant, available at Roma Tre University facilities, with direct current (DC)-powered aluminium targets and a radio frequency (RF)-powered capacitive coupled sample holder capable of inducing bias to conductive and dielectric substrates. More details on the deposition geometry are reported in previous papers [3,4]. Nitrogen was put into the chamber to produce aluminium nitride by reactive sputtering. The aluminium nitride thin films were deposited on ≈290 μm thick rectangular (20 mm long and 7 mm wide) single-crystal (100) silicon wafers with an initial average roughness (Ra) of ≈10 nm. Prior to sputtering, silicon substrates were cleaned with acetone and ethanol for five minutes in each. Samples were mounted on a sample holder that was placed at a distance of 80 mm from the target. These substrates were subsequently sputter-etched in argon plasma for 10 min in order to remove the surface oxide layer. Four samples were deposited with an AlN coating using 180 W DC on the 200 mm Al cathode, a gas mixture of 25% argon and 75% nitrogen, a process pressure of 1×10^{-4} mbar, and a base pressure lower than

6×10^{-6} mbar. With these deposition parameters, the cathode was fully poisoned, ensuring a perfect stoichiometry of the produced coatings; the low power density on the aluminium targets ensured no arcing and thus no particle production. The four samples differed for the applied negative bias voltage to the substrate during growth and the optional presence of a 50 nm thin pure aluminium bond layer; the samples with or without the bond layer were coated with floating potential (around 30 V from the plasma) or 100 V RF applied potential. The maximum temperature developed during deposition and ion etching was measured with a temperature measurement strip gauge. These strip gauges were mounted onto the aluminum sample holder which held the sample on other side. At the end of deposition, a maximum temperature of ≈150 °C was recorded. To facilitate reading, the produced coatings were investigated by dividing into two groups: one without the bond layer under a floating bias of 30 V and bias voltage of 100 V, named AlN-30V and AlN-100V, respectively; in the second group was an AlN coating under a bias voltage of 30 V and 100 V with an aluminium bond layer, named AlN/Al-30V and AlN/Al-100V, respectively.

After deposition, the wafer curvature method (extended Stoney formula) was used for average residual stress measurement, in which the biaxial state of stress was taken into consideration [9,10]. The choice of (100) silicon, with its more complex stiffness with respect to (111) silicon, was due to the need to have perfectly rectangular strips ((111) silicon tends toward triangular cleavage). Curvature profiles were measured on the coating side. As widely known from the literature, a concave surface curvature corresponds to compressive residual stress in the characterized coatings. Curvature was then measured using a state-of-the-art optical profilometer (Leica DCM-3D, Bannockburn, IL, USA) in accordance with the standard procedure CEN/TS 1071-1 [11]. An optical profilometer enables accurate and reproducible curvature measurements with z-height resolution of ≈0.6 nm. A two-dimensional (2D) coating surface profile with up to 8 mm scan length was constructed with extended topography in confocal mode. The radius of curvature was measured through circle fitting of the 2D profile, and residual stresses were calculated by the modified Stoney equation:

$$\sigma = M_{(100)} \frac{t_s^2}{6t_c} \frac{1}{R} \tag{1}$$

where t_s and t_c are the thickness of the substrate and coating, respectively, and M (100) = $\frac{E_s}{1-\vartheta_s}$ is the biaxial modulus of single-crystal silicon (100), which is 180.4 GPa [12]. In the equation above, E_s and ϑ_s are the elastic modulus and Poisson's ratio of the substrate, respectively. The thickness of the coating was measured with an optical profilometer (step height method) and verified with focused ion beam (FIB) FEI Helios NanoLab 600 Dualbeam FIB/SEM (Thermo Scientific, Hillsboro, OR, USA) cross-section analysis.

The through-thickness gradient of residual stress was measured by using a novel incremental micro-ring-core profiling method [13–15], which consists of controlled material removal by focused ion beam (FIB) microscopy, coupled with SEM high-resolution imaging, digital image correlation (DIC) for relaxation strain analysis, and finite-element (FEM) calculation of the residual stress depth gradient. As described in a recent paper, this new method allows for nanoscale depth profiling [15] of the residual stress in thin films. In this paper, this method was applied to the coatings with an Al interlayer (both deposited at −30 and −100 V bias voltage) in order to investigate the residual stress gradient in proximity of the AlN/Al interface. The same method also allowed for the estimation of the average residual stress in the film, and a comparison with the results from curvature analysis was made.

The elastic modulus and hardness of coatings were measured by nanoindentation testing in accordance with the ISO 14577 standard (Keysight G200 Nano Indenter®, 0.05 s^{-1} constant strain rate, maximum indentation depth 1000 nm) by employing the continuous stiffness method (CSM), which allows to calculate the hardness (H) and elastic modulus (E) profiles versus penetration depth (it is the superposition of a oscillating load to the linear increment in the indenter depth). For all nanoindentation experiments, the frame compliance and indenter tip area function were calibrated on a standard fused silica block and following the procedure suggested by Oliver and Pharr [16]. As the coating had a thickness in the order of ≈700 nm, the most reliable values of hardness and modulus were obtained by interpolation of the CSM hardness/modulus versus depth profiles and the calculation of the interpolated average H and E values in the depth range of 60–70 nm. In this way, the effects coming from the substrate's influences are reduced (this is especially true for hardness, while the elastic modulus is always influenced by the substrate compliance).

Adhesion of coatings was assessed by equipping the nanoindenter with a cube-corner tip and making load-controlled indentations in order to induce coating delamination, using a maximum load of 200 mN. Nine indents on each sample were made and at least five indents were observed by scanning electron microscopy (SEM). The interfacial adhesion of coatings under different residual stress conditions were evaluated using the model described by den-Toonder et al. [17], through microscale observation of indentation-induced failures. Originally, Thouless [18] proposed this model for interfacial adhesion through the observation of triangular coating spallation failures with scratch tests. Later, den-Toonder et al. [17] modified this model to take into account the curved geometry of the delaminated segment and residual stresses, which usually produces failures in residual stressed coatings under nanoindentation. In this work, this model was used to quantify the interfacial adhesion of the films.

3. Results and Discussion

3.1. Residual Stress Measurement

An isometric view of a typical 2D profile of silicon wafer curvature measured with an optical profilometer is shown in Figure 1; curvatures for other samples are not shown here in the interest of simplicity.

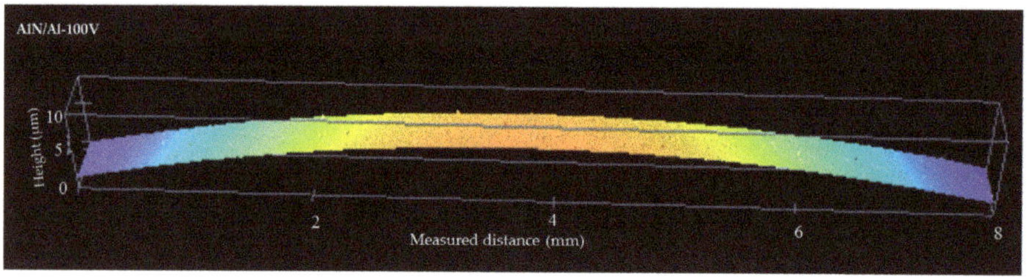

Figure 1. Example of optical profilometer-measured radius of curvature of the coatings after deposition.

Coating thickness and microstructure were analyzed with FIB and transmission Electron Microscopy (TEM) section analysis (Figure 2), while residual stress was calculated using the curvature method, as reported in Table 1. As clearly visible in Figure 2, the presence of an Al bond layer does not change the microstructure of the AlN layers, which are made of very fine columnar nanostructured grains.

Table 1. Description of coatings' deposition parameters, measured overall coating thickness, and compressive residual stress with the wafer curvature method.

Description of Coatings	Bias Voltage (V)	Bond Layer (Yes/No)	Thickness (μm)	Average Residual Stress (GPa)
AlN-30V	−30	No	0.75 ± 0.05	−1.2 ± 0.2
AlN/Al-30V	−30	Yes	0.76 ± 0.02	−1.5 ± 0.2
AlN-100V	−100	No	0.72 ± 0.03	−3.5 ± 0.2
AlN/Al-100V	−100	Yes	0.71 ± 0.04	−3.9 ± 0.2

The average residual stress values are shown in Figure 3, as calculated by Stoney and FIB-DIC methods. For the two samples where two techniques are used, there is good agreement between the data (i.e., the discrepancy is within the standard deviations), as also reported in previous literature.

On comparing the coatings in the first and second groups, significantly higher residual stress was found under polarization of the substrate with negative potential of 100 V (AlN-100V coatings), compared to the substrate under polarization of 30 V (AlN-30V coating). This is an expected result, and the increase of the average compressive residual stress for all coatings that were deposited at −100 V bias voltage can be attributed to an atom peening effect given by the increased energy of the vapor flux. In case of the AlN coatings on silicon, the thermal component of the residual stress can be considered as negligible with respect to the atomic peening component, since the thermal expansion coefficients of the two materials (AlN and silicon) are similar.

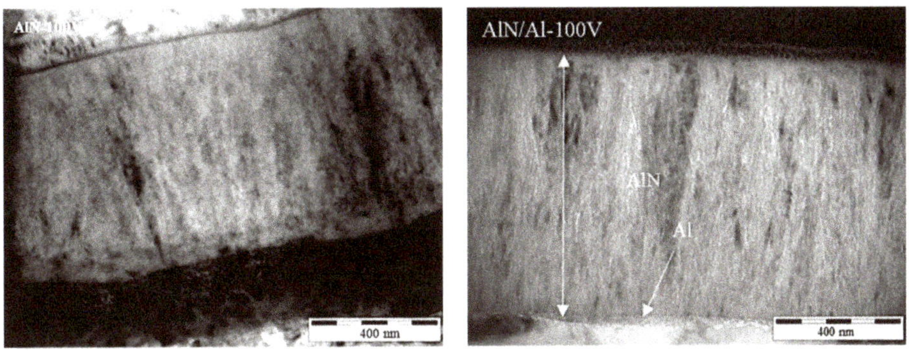

(a) (b)

Figure 2. TEM observations (of focused ion beam cross sections) of the produced coatings for coating thickness measurement: (**a**) AlN-100V; (**b**) AlN/Al-100V.

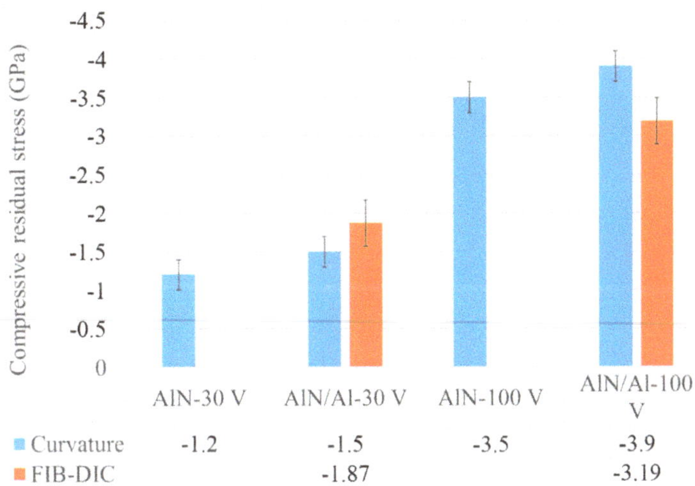

Figure 3. Plot of average compressive residual stress (curvature measurements) and focused ion beam–digital image correlation (FIB-DIC) in all the produced coatings.

On the other hand, in the AlN coating with an Al bond layer, the thermal stress component has contributed significantly to an increase in the overall residual stress in the AlN/Al coating system, as shown in Figure 3 (under the assumption that the atomic peening component remains the same).

Castanho et al. [19] observed a similar phenomenon in magnetron sputtered multilayer coatings, whereas the number of aluminum interlayers was increased, more compressive stresses were induced in the top layer of AlTiN. In fact, the contribution of thermal stresses was analytically evaluated to be ≈+340 MPa by using the thermal expansion of the substrate (Si, 2.3×10^{-6} °C^{-1}) and bond layer (Al, 22.8×10^{-6} °C^{-1}) reported in the literature [7]. The aluminum bond layer during cooling from the deposition temperature attempted to contract more than the silicon substrate and the AlN coating. However, at the interfaces, the aluminum cannot contract fully due to being constrained by the silicon substrate and AlN top layer, and as a result, the aluminum bond layer is in tension, which causes additional compression residual stresses in the AlN coating and silicon substrate. By taking these thermal stress contributions into account, the residual stress variation between the coatings with and without the bond layer as reported in Figure 3 is largely explained.

Furthermore, the FIB-DIC analysis of residual stress depth profiles in the AlN layers is a further confirmation of the above discussions. In fact, in both cases (AlN/Al with −30 V and −100 V bias), we do observe a strong surface compressive residual stress that goes towards tensile when approaching the AlN/Al interface. In the case of the AlN/Al-100V system, we even observed a mild state of tensile stress in the AlN layer in the vicinity of the interface. For this sample, the presence of tensile stress when approaching the interface is a further explanation for the extensive delamination that was observed during the indentation.

Therefore, the adoption of an additional Al interlayer, in this specific case, gave rise to an increased compressive stress in the top layer (AlN), and at the same time, a stronger stress gradient and an additional tensile residual stress at the interface (Figure 4c).

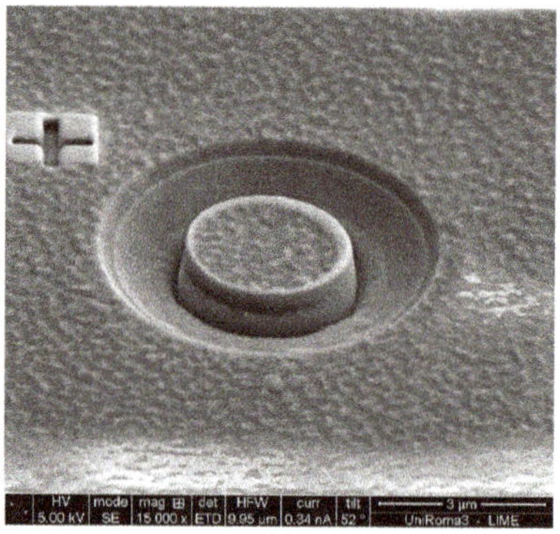

(**a**)

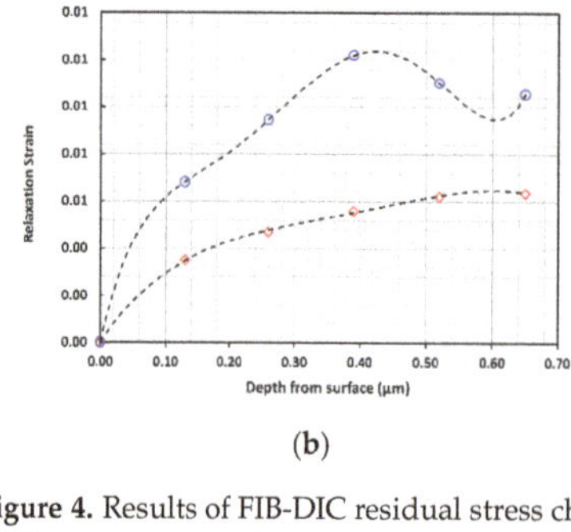

(**b**)

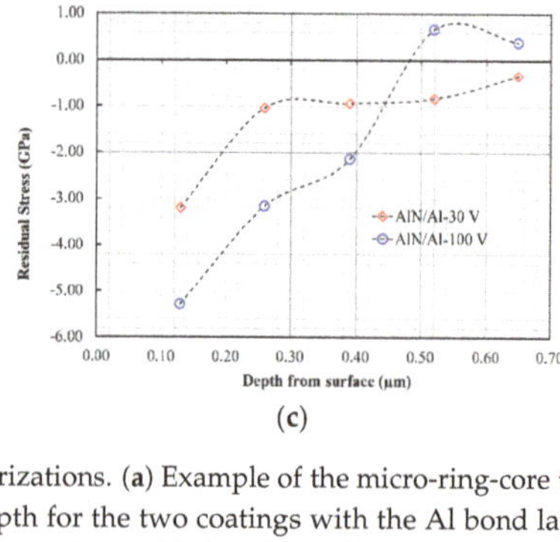

(**c**)

Figure 4. Results of FIB-DIC residual stress characterizations. (**a**) Example of the micro-ring-core test; (**b**) Representative relaxation strains vs. milling depth for the two coatings with the Al bond layer; (**c**) Representative residual stress profiles for the two coatings with the Al bond layer (AlN/Al with −30 V and −100 V bias), obtained by using the FIB-DIC depth profiling method described in [15].

3.2. Nanoindentation Characterizations

3.2.1. Hardness and Elastic Modulus

The hardness and elastic modulus of all the produced coatings were measured with the Berkovich indenter, and results are summarized in Figure 5.

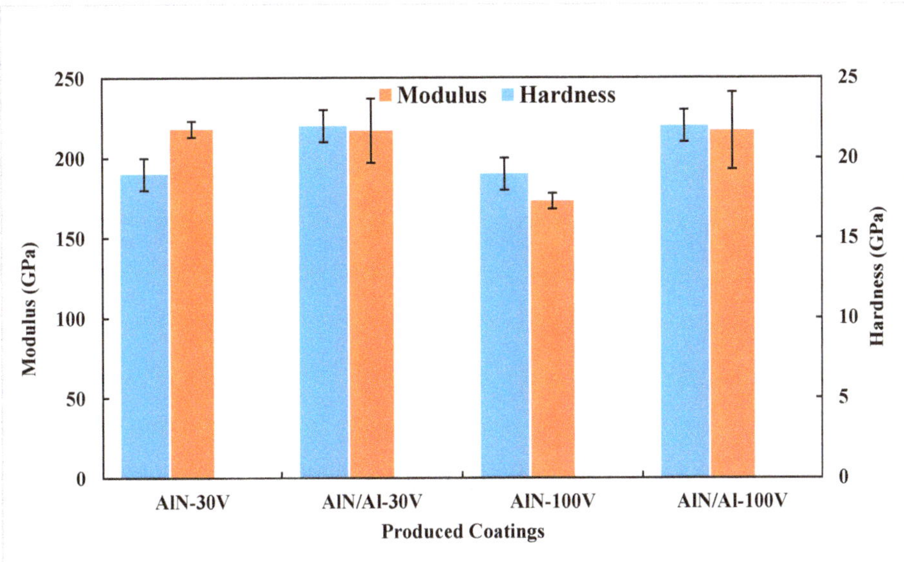

Figure 5. Results of nanoindentation hardness and elastic modulus.

There is a minor difference in the measurements between AlN coatings with and without the bond layer. Coatings made with the bond layer have higher standard deviations that those without it. The major source of deviation in nanoindentation hardness and modulus is roughness of the coating surface, and it is possible to see from Figure 6 that the surfaces of the coatings AlN-30V and AlN/Al-30V are different.

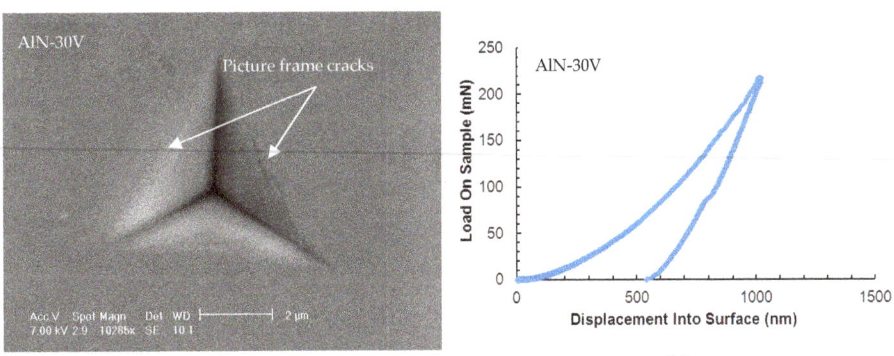

(**a**) (**b**)

Figure 6. *Cont.*

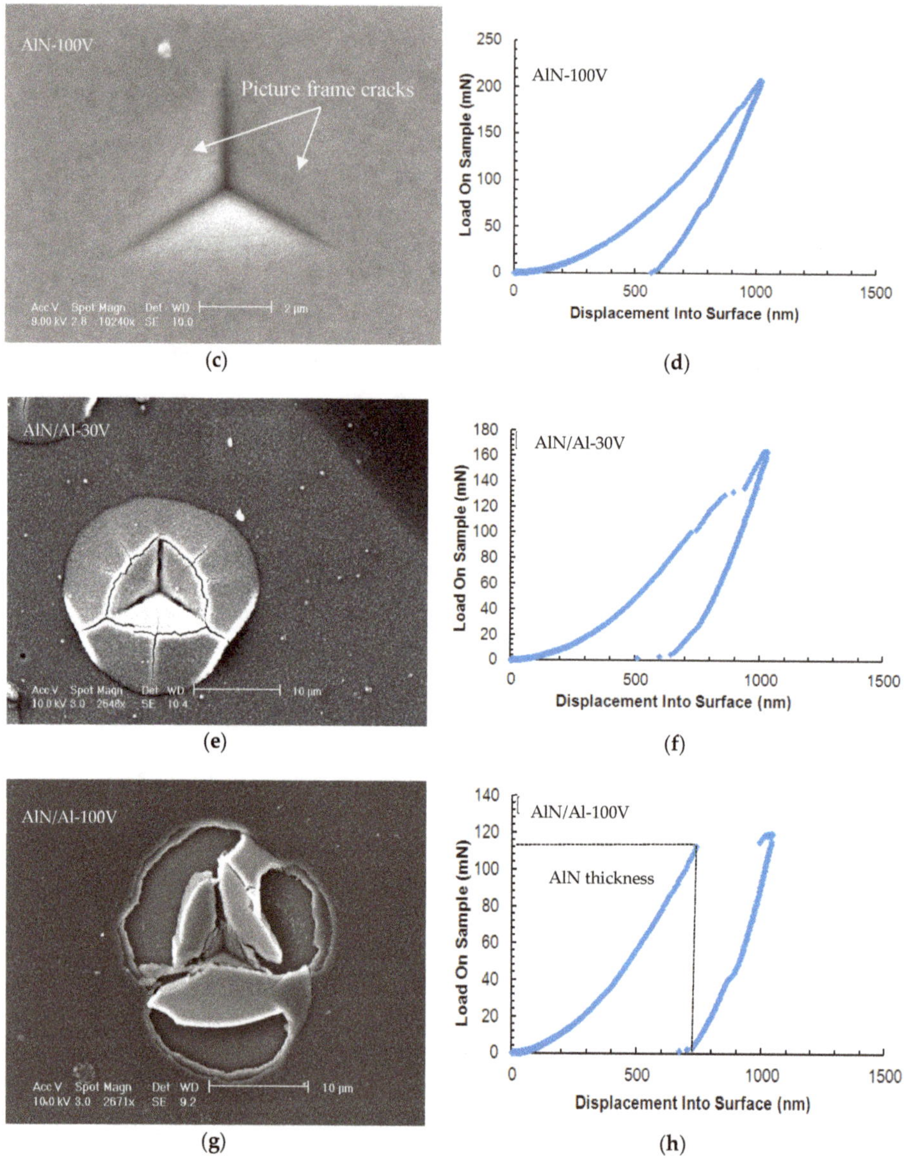

Figure 6. The SEM micrographs of indentations (**left**) made with the Berkovich indenter and the associated load–displacement curves (**right**). (**a**,**b**) AlN-30V, (**c**,**d**) AlN-100V, (**e**,**f**) AlN/Al-30V, (**g**,**h**) AlN/Al-100V.

The nanoindentation results further demonstrate that the only (and major) difference among the coatings is the residual stress and its distribution. Therefore, the main difference in adhesion could only be attributed to the different residual stress distribution.

The subsequent sections summarize the results of coating/substrate interfacial adhesion, failure modes, and their effect analysis.

3.2.2. Quantitative Evaluation of Coatings' Interfacial Adhesion

The interfacial adhesion energy of the coatings was investigated by analyzing microscale delamination failures produced under Berkovich and cube-corner indenter tips. At first, nanoindentations were made with standard Berkovich indentation tip. Figure 5 shows the load–displacement curves and the associated SEM micrograph of an indent in the four cases under consideration.

As clearly visible from Figure 6, the coating/substrate adhesion decreases dramatically for the two coatings with an Al layer.

The discontinuous cracks formed in the indents of the AlN-30V and AlN-100V coatings (as indicated by the picture-frame cracks shown in Figure 6) seem to be due to tensile stresses induced by the dragging of the part of the coating around the indenter from the bulk of the coatings. An extensive analysis of this kind of cracks is reported in [20], where it is explained that they are related to the E/H (elastic modulus to hardness) ratio mismatch between the coating and substrate. The discontinuity in displacements during the unloading segment is related to the reverse phase transformation in the silicon substrate [21].

Only the coatings with the Al bond layer, AlN/Al-30V and AlN/Al-100V, presented extensive buckling and delamination with the Berkovich indenter at 140 mN and 120 mN indentation loads, respectively, as shown in the right column of Figure 5. The load–displacement curve in the coating AlN/Al-100V at the thickness of ≈750 nm exhibited a displacement burst (horizontal step) and confirms the through-thickness cracking leads to the interfacial detachment of the coating. The metallic film (bond layer) is likely to undergo plastic deformation and these deformations could be significantly larger than the film thickness [22]. From these observations, it was clear that the coating AlN/Al-100V has the poorest adhesion, as it delaminated completely at the lower load and it also had higher residual stress in comparison with the AlN/Al-30V coating.

After these tests, all the four coatings were tested again with a cube corner indenter [23] to produce fractures. The coatings (AlN-30V and AlN-100V) without the bond layer did not delaminate up to the substrate (Figure 7) and only showed cracks within the film thickness.

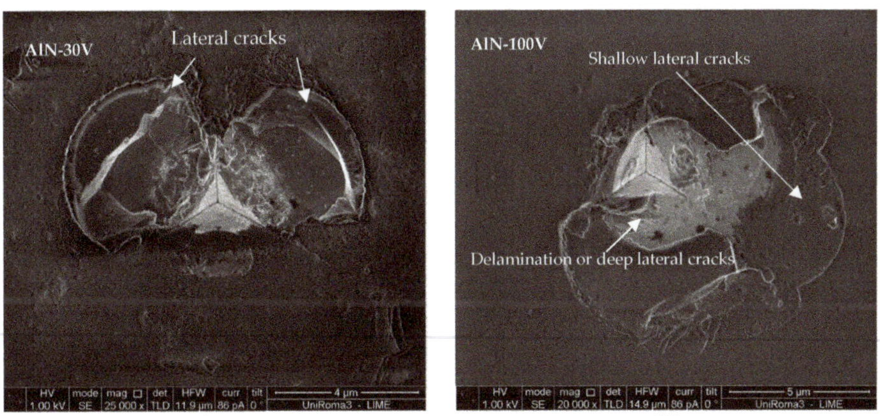

(a) (b)

Figure 7. The SEM micrographs of indentations made with a cube-corner indenter on the coatings: (**a**) AlN-30V and (**b**) AlN-100V (at different magnifications).

The SEM indentation micrographs of the coatings AlN-30V and AlN-100V clearly shows that the central region is shallow to various depths compared to the mean level of the coating (Figure 7) and the cracks' morphology is similar to the usually observed lateral cracks [24]. The propagation of cracks through the formation of steps, followed by turning upward to form chips, is confirmation of lateral cracks, and crack propagation is similar in the two coatings' micrographs. It is possible that higher residual stress magnitude caused the chipping (possible delamination) failure, seemingly due to weaker adhesion in the AlN-100V coating. However, for the AlN-30V coating, having good adhesion, the indenter caused lateral cracks without delamination. The possibility of change in the crack pattern with a sharp cube-corner indenter is expected as a result of the substrate effect.

For this lateral cracking in the coatings AlN-30V and AlN-100V, a fracture toughness value of 1.8 and 2.3 MPa·m$^{0.5}$ was estimated with the den-Toonder model (Equation (1)) [17], which was originally developed by Antis et al. [25] to investigate the fracture toughness of bulk materials. The inputs of this model are the crack load (80 and 100 mN for the coatings AlN-30V and AlN-100V, respectively), the length of lateral cracks, the crack depth which is equal to half the coating thickness, and the measured residual stresses in the coatings. It has to be underlined that such values should be only considered as providing a rough estimation of a coating's fracture toughness, since there are many factors affecting the reliability of such indentation-based methods, as discussed in recent papers [26]. Nonetheless, the estimated fracture toughness values of the coatings are very similar and in the range of the toughness (1.5–5 MPa·m$^{0.5}$) of ceramic coatings reported in the literature. This estimation has an error of 0.2 MPa·m$^{0.5}$ (normally, the error is much higher for this kind of measurement, at around 30%); this low error value comes from the repeatability of the results. It was expected that higher stresses under higher bias voltages were to be correlated with higher toughness of the coating, but this was not the case. The measurements show that the higher load needed to crack the coating AlN-100V is due to the compressive intrinsic stresses, and once this effect is taken into account, there are no major toughness difference between the two coatings.

For the coatings with the Al bond layer, the cube-corner indenter achieved full coating delamination from the substrate, and the den-Toonder model [17] could be used to quantify the interfacial adhesion energy. According to this model, the indenter only has the role of crack initiation, and residual stresses are the driving force for delamination and chipping. The inputs of this model are the measured coating thickness, elastic modulus, residual stresses, and chipped coating segment geometric information, as elaborated in Figure 8a.

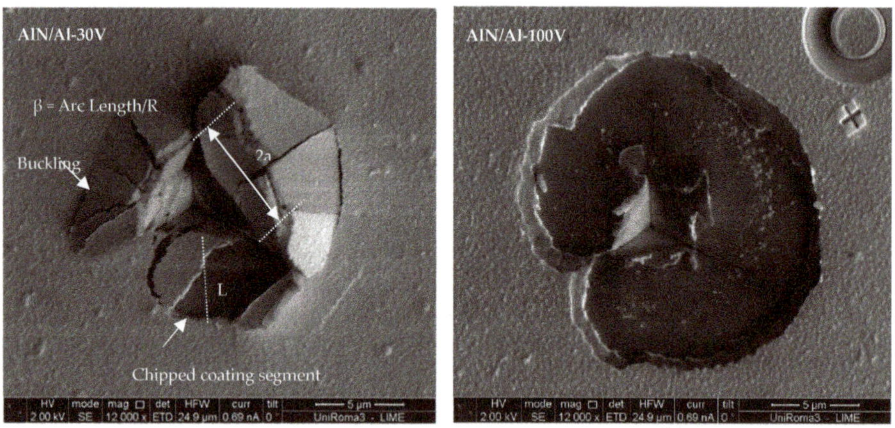

(a) (b)

Figure 8. The SEM micrographs of indentation-induced failures for adhesion measurement of (**a**) AlN/Al-30V and (**b**) AlN/Al-100V coatings.

Coating segments formed chips, which can be clearly seen in the micrograph of the AlN/Al-30V coating, in a semicircle with an easily identifiable radius, and the chipped region has similar schematics for which den-Toonder developed the model (Figure 8). With comparison of the microscale delamination failure mode (semicircular) in Figure 7, it is also possible to see that the circular chipping in the AlN/Al-100V coating was larger than that in the AlN/Al-30V coating. The AlN/Al-30V coating had lower residual stress in comparison with the AlN/Al-100V coating, and having higher residual stresses resulted in different delamination failures; however, the mode of delamination in almost all of the indents was circular (Figure 9).

The model of den Toonder et al. [17] (Equation (10)) resulted in the interfacial adhesion energy of 5 and 4 J/m^2 for the AlN/Al-30V and AlN/Al-100V coatings, respectively. The estimations have an error of $\pm 5\%$. The calculated adhesion energy is the practical interfacial fracture energy and it also includes the elastic strain energy associated with the release of residual stresses. The interfacial fracture energy or the adhesion energy results are in accordance with the qualitative comparison of indentation-induced failures produced with the cube-corner indenter. In addition, the adhesive failure modes of the AlN/Al-30V and AlN/Al-100V coatings were compared with the literature, and were in good agreement with those given by Bull [27]. The results indicate that the adhesion of the coating has been deteriorated, in this specific case, due to the presence of the aluminum bond layer. As the AlN-30V and AlN/Al-30V coatings have a difference in compressive residual stress of only +340 MPa, we can expect that the Al bond layer is in a tensile stress state with a similar magnitude.

This assumption can be justified by considering the additive nature of residual strains in multilayer systems. The bond layer, being ductile, cannot store elastic energy across the interface, and tensile residual stresses are also present in it (as also observed by FIB-DIC stress profiles). The tension coupled with deformation during indentation in the Al bond layer leads to interfacial delamination. In most of the studies available in the literature, AlN was directly coated onto a silicon substrate; nonetheless, there are a few studies that used a thin layer of Al to improve the bonding between the AlN and Si substrate, even though the adhesion was not investigated [28].

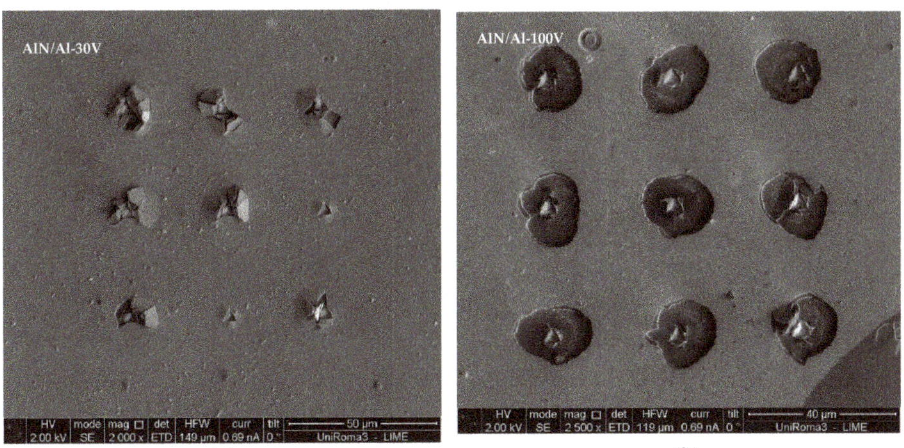

(a) (b)

Figure 9. The SEM micrographs of indentation-induced failures of all the indents made for adhesion measurement in (**a**) AlN/Al-30V and (**b**) AlN/Al-100V coatings (at different magnifications).

In this work, we have shown that the role of the residual stress in the bond layer is significant, because even a ≈50 nm thin layer is capable of causing a break by bulging the ≈700 nm thick upper layer, decreasing the functional adhesion of the coating under nanoindentation (Figure 9). Therefore, we demonstrate here that the residual stress state in the Al bond layer can have a primary and relevant effect on the coating's adhesion, more than the residual stress and hardness of the AlN top layer.

Also, with bond layer being ductile, one would expect an increase of the interfacial adhesion energy because of the decrease of stored elastic energy across the interface. However, as the indentation depth exceeds the top layer's thickness (AlN), a step in the load–displacement curve occurs (see right column of Figure 6) in the AlN/Al-30V and AlN/Al-100V coatings. As a result, soft bond layers, such as of pure aluminum and copper, are not adequate to be a proper coating base, especially if the substrate is hard. This is the reason why only bond layers made of Ti, Cr, Mo, or Nb are effective on the hard substrate for the smooth transition of properties between the hard coating and substrate [29,30]. Comparison of coating delamination behavior under different residual stress conditions suggested (Figure 8) that both the mode I and mode II load components in the buckled and unbuckled states are operative. The schematic of both the buckled and unbuckled failure modes are given in Figure 10 for reference.

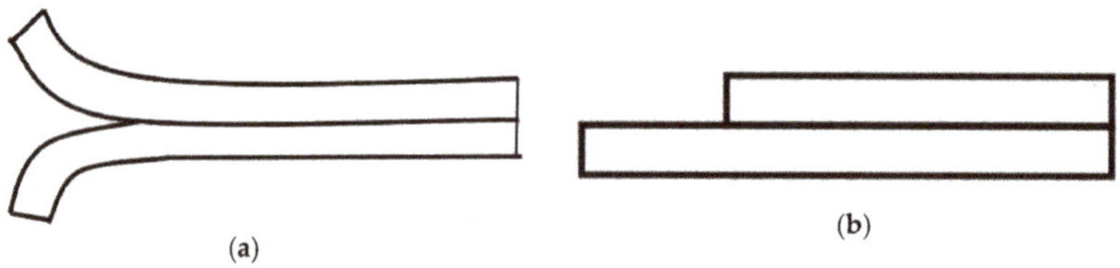

Figure 10. Schematic of interfacial crack initiation modes: (**a**) mode I failure; (**b**) mode II failure.

It can be seen in the micrograph of Figure 8b that the AlN/Al-100V coating shows shear mode II and the failure can be characterized as catastrophic with the highest delamination; however, in the AlN/Al-30V coating, a mixture of peeling (mode I) and shear (mode II) delamination failures were observed as a result of tensile stress in the aluminum coating layer and compressive stresses in the AlN layer. It is possible that at about the critical indenter penetration, the higher compressive stresses of indentation resulted in the buckling of the soft thin film of aluminum, which means that the interfacial adhesion was not too high. The buckling failure is more likely than the shear-induced delamination when the coating is relatively tough compared to the interface. In order to confirm whether the coating delaminates from the AlN/bond layer interface or from the bond layer/substrate interface, energy dispersive spectroscopy (EDS) composition maps were constructed (Figure 11).

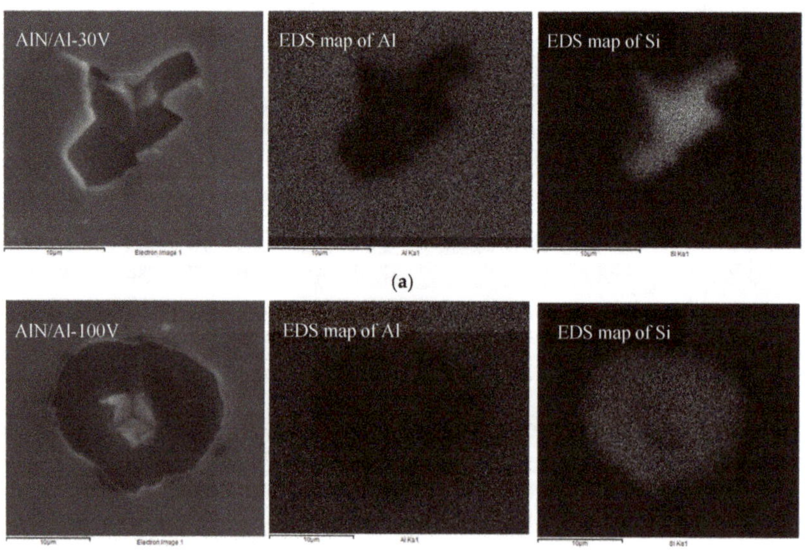

Figure 11. The energy dispersive spectroscopy (EDS) maps of cube-corner produced indentation marks in coatings and the corresponding mapping images of the Al and Si elements: (**a**) AlN/Al-30V; (**b**) AlN/Al-100V.

The EDS map of aluminum in the second column of Figure 11 shows that aluminum is present in the AlN layer; however, it is not present where the coating has been delaminated. Similarly, the silicon EDS map in the third column from the left shows that silicon is not present in the AlN coating and silicon substrate is exposed after delamination. It is noticeable that silicon substrate is visible and the bond layer is missing in the EDS map, even though the bond layer thickness was measured and confirmed with FIB-TEM cross-section analysis (shown and indicated in Figure 2). This shows that delamination failures are from the bond layer and substrate interface, while the bonding between AlN and Al is perfect. The results indicate that the adhesion of coating has been deteriorated due to the aluminum bond layer and the surface fracture is attributable to the interface between the aluminum and silicon. Using this information and the measured interfacial delamination energy (G_{int}) of 5 and 4 J/m^2 for the AlN/Al-30V and AlN/Al-100V coatings, respectively, the interface fracture toughness (K_{int}) defined by Suo and Hutchinson [31] may be estimated using Equation (2).

$$K_{int} = \sqrt{G_{int} E_{int}} \tag{2}$$

where E_{int} is the interface's elastic modulus, given by Equation (3).

$$\frac{1}{E_{int}} = \frac{1}{2}\left[\frac{1}{E} + \frac{1}{E_s}\right] \tag{3}$$

Using E as the coating modulus and E_s as the substrate modulus, the interface elastic modulus (E_{int}) was found to be 100 GPa. The estimated interface fracture toughness of 0.71 and 0.65 $MPa{\cdot}m^{0.5}$ for the AlN/Al-30V and AlN/Al-100V coatings, respectively, clearly indicate why these coatings delaminated instead of cracking, as the resulting interfacial fracture toughness is significantly lower than the cohesive fracture toughness of the AlN-30V and AlN-100V coatings.

There are technological implications of this study's results for the design and production of coatings on hard substrates, in particular on silicon substrates, for various applications. For example, aluminum nitride and other ceramic coatings are of industrial interest as next-generation materials for sensor applications. Ionescu et al. [32] indicated the real possibilities of manufacturing sensors embedded on the monitored components of machine tools [33]. The strength of silicon wafers used in solar cell and microelectronics applications could be improved with hard coatings [34]. Also, the microscale-produced failure modes and their effect analysis is additive to contact-induced coating failures, which could help to validate the coating adhesion models. In this context, the experimental characterization adds new information to the literature that may help to interpret the contact-induced failures in coatings under tribological situations, by focusing in particular on the role of the residual stress distributions for the case of metal/ceramic multilayer systems.

4. Conclusions

The effects of residual stress distribution on the interfacial adhesion energy of coatings were quantitatively investigated and related to the observed micromechanical behavior. The following conclusions could be drawn from this work:

1. An increase in the average compressive residual stress in AlN coatings on the silicon substrate was measured with an increase in bias voltage and coatings deposited with the Al bond layer. Higher bias voltage accelerates the gas ions toward the growing coating and the additional energy caused the enhancement of residual stresses. It is also possible that the different bias conditions affect the interface between Si and AlN, in addition to increasing the compressive residual stresses. The residual stress was further increased by adding an Al bond layer, as a result of its high thermal expansion coefficient in comparison with AlN and the silicon substrate. As a consequence, tensile thermal stresses during cooling were established in the Al, which caused additional compressive residual stresses in the AlN/Al coatings.

2. FIB-DIC residual stress depth profiles showed that for the AlN/Al coatings, a strong residual stress gradient was present in the films, being the maximum compressive residual stress on the top surface, then decreasing towards the interface; a mild tensile stress was observed even for the AlN/Al-100V sample, which was the one showing more extended delamination.
3. The coatings deposited without the Al bond layer showed significantly better adhesion in comparison to those with the bond layer, even though both groups of coatings had differences in compressive residual stresses, equivalent to thermal stresses in the bond layer. Adhesive failures could not be produced in the coatings without a bond layer (AlN-30V and AlN-100V), even with a cube-corner indenter, which means that the interface adhesion was higher in comparison with the coatings deposited with the Al bond layer. The fracture toughness corresponding to lateral cracks was measured to be 2.3 and 1.8 MPa $m^{0.5}$ for the coatings AlN-30V and AlN-100V, having residual stresses of 1.2 and 3.5 GPa, respectively. The observed behavior could be related to more homogenous compressive residual stress distribution in the AlN top layer.
4. The adhesion energy for interfacial separation was found to be 5 and 4 J/m^2 for the coatings AlN/Al-30V and AlN/Al-100V, having an average residual stress of 1.5 and 3.9 GPa, respectively, with a corresponding interfacial fracture toughness of 0.71 and 0.65 $MPa{\cdot}m^{0.5}$, which was significantly lower than the estimated fracture toughness of the AlN material, thus explaining why delamination occurs for AlN/Al systems. The tensile residual stress on the bond layer, coupled with its plastic deformation during indentation, leads to interface delamination.
5. These results suggested that in order to be beneficial for adhesion, a metallic bond layer should possess good hardness, a small difference in the thermal expansion coefficient, and a comparable elastic modulus with both the substrate and top coating. Refractory metals are a good prospect for the design of improved multilayer coating systems with optimal performance.

Author Contributions: R.A. and M.R. performed the experimental activities and wrote most sections of the paper. M.S., M.I.K., and E.B. critically reviewed the experimental data and made revisions to the paper.

Funding: M.S. efforts received funding from the European Union, within the large collaborative project OYSTER, Grant Agreement No. 760827, www.oyster-project.eu.

Acknowledgments: The authors would like to thank Daniele de Felicis for his cooperation in focused ion beam imaging for coating thickness measurements. All the measurements and characterization activities were carried out at the Inter-Departmental Laboratory of Electron Microscopy (LIME), University of ROMA TRE (http://www.lime.uniroma3.it), Italy.

Conflicts of Interest: The authors declare no conflict of interest.

References

1. Bian, Y.; Liu, M.; Ke, G.; Chen, Y.; DiBattista, J.; Chan, E.; Yang, Y. Aluminum nitride thin film growth and applications for heat dissipation. *Surf. Coat. Technol.* **2015**, *267*, 65–69. [CrossRef]
2. Abadias, G.; Chason, E.; Keckes, J.; Sebastiani, M.; Thompson, G.B.; Barthel, E.; Doll, G.L.; Murray, C.E.; Stoessel, C.H.; Martinu, L. Stress in thin films and coatings: Current status, challenges, and prospects. *J. Vac. Sci. Technol. A Vac. Surf. Films* **2018**, *36*, 020801. [CrossRef]
3. Ali, R.; Sebastiani, M.; Bemporad, E. Influence of Ti–TiN multilayer PVD-coatings design on residual stresses and adhesion. *Mater. Des.* **2015**, *75*, 47–56. [CrossRef]
4. Renzelli, M.; Mughal, M.Z.; Sebastiani, M.; Bemporad, E. Design, fabrication and characterization of multilayer Cr-CrN thin coatings with tailored residual stress profiles. *Mater. Des.* **2016**, *112*, 162–171. [CrossRef]
5. Kleinbichler, A.; Pfeifenberger, M.J.; Zechner, J.; Moody, N.R.; Bahr, D.F.; Cordill, M.J. New Insights into Nanoindentation-Based Adhesion Testing. *JOM* **2017**, *69*, 2237–2245. [CrossRef]
6. Kriese, M.D.; Gerberich, W.W.; Moody, N.R. Quantitative adhesion measures of multilayer films: Part II. Indentation of W/Cu, W/W, Cr/W. *J. Mater. Res.* **1999**, *14*, 3019–3026. [CrossRef]
7. Khanna, V. Adhesion–delamination phenomena at the surfaces and interfaces in microelectronics and MEMS structures and packaged devices. *J. Phys. D Appl. Phys.* **2010**, *44*, 034004. [CrossRef]

8. Liu, D.; Pallon, L.K.H.; Pourrahimi, A.M.; Zhang, P.; Diaz, A.; Holler, M.; Schneider, K.; Olsson, R.T.; Hedenqvist, M.S.; Yu, S.; et al. Cavitation in strained polyethylene/aluminium oxide nanocomposites. *Eur. Polym. J.* **2017**, *87*, 255–265. [CrossRef]
9. Stoney, G.G. The tension of metallic films deposited by electrolysis. *Proc. R. Soc. Lond. Ser. A Contain. Pap. Math. Phys. Character* **1909**, *82*, 172–175. [CrossRef]
10. Janssen, G.C.A.M.; Abdalla, M.M.; van Keulen, F.; Pujada, B.R.; van Venrooy, B. Celebrating the 100th anniversary of the Stoney equation for film stress: Developments from polycrystalline steel strips to single crystal silicon wafers. *Thin Solid Films* **2009**, *517*, 1858–1867. [CrossRef]
11. Advanced technical ceramics. Methods of test for ceramic coatings. Determination of internal stress by the Stoney formula. In *BS DD CEN/TS 1071*; BSI: London, UK, 2005.
12. Daniel, R.; Martinschitz, K.; Keckes, J.; Mitterer, C. The origin of stresses in magnetron-sputtered thin films with zone T structures. *Acta Mater.* **2010**, *58*, 2621–2633. [CrossRef]
13. Korsunsky, A.M.; Sebastiani, M.; Bemporad, E. Focused ion beam ring drilling for residual stress evaluation. *Mater. Lett.* **2009**, *63*, 1961–1963. [CrossRef]
14. Bemporad, E.; Brisotto, M.; Depero, L.; Gelfi, M.; Korsunsky, A.; Lunt, A.; Sebastiani, M. A critical comparison between XRD and FIB residual stressmeasurement techniques in thin films. *Thin Solid Films* **2014**, *572*, 224–231. [CrossRef]
15. Korsunsky, A.M.; Salvati, E.; Lunt, A.G.J.; Sui, T.; Mughal, M.Z.; Daniel, R.; Keckes, J.; Bemporad, E.; Sebastiani, M. Nanoscale residual stress depth profiling by Focused Ion Beam milling and eigenstrain analysis. *Mater. Des.* **2018**, *145*, 55–64. [CrossRef]
16. Oliver, W.C.; Pharr, G.M. An improved technique for determining hardness and elastic modulus using load and displacement sensing indentation experiments. *J. Mater. Res.* **1992**, *7*, 1564–1583. [CrossRef]
17. Toonder, J.D.; Malzbender, J.; With, G.D.; Balkenende, R. Fracture Toughness and Adhesion Energy of Sol-gel Coatings on Glass. *J. Mater. Res.* **2002**, *17*, 224–233. [CrossRef]
18. Thouless, M.D. An analysis of spalling in the microscratch test. *Eng. Fract. Mech.* **1998**, *61*, 75–81. [CrossRef]
19. Castanho, J.; Vieira, M. Effect of ductile layers in mechanical behaviour of TiAlN thin coatings. *J. Mater. Process. Technol.* **2003**, *143*, 352–357. [CrossRef]
20. Chen, J.; Bull, S. Indentation Fracture and Toughness Assessment for Thin Optical Coatings on Glass. *J. Phys. D Appl. Phys.* **2007**, *40*, 5401. [CrossRef]
21. Pharr, G.M.; Oliver, W.C.; Harding, D.S. New evidence for a pressure-induced phase transformation during the indentation of silicon. *J. Mater. Res.* **1991**, *6*, 1129–1130. [CrossRef]
22. Haberl, B.; Bradby, J.; Swain, M.; Williams, J.; Munroe, P. Phase transformations induced in relaxed amorphous silicon by indentation at room temperature. *Appl. Phys. Lett.* **2004**, *85*, 5559–5561. [CrossRef]
23. Ghidelli, M.; Sebastiani, M.; Johanns, K.E.; Pharr, G.M. Effects of indenter angle on micro-scale fracture toughness measurement by pillar splitting. *J. Am. Ceram. Soc.* **2017**, *100*, 5731–5738. [CrossRef]
24. Morris, D.J.; Cook, R.F. Indentation fracture of low-dielectric constant films: Part I. Experiments and observations. *J. Mater. Res.* **2008**, *23*, 2429–2442. [CrossRef]
25. Anstis, G.; Chantikul, P.; Lawn, B.R.; Marshall, D. A critical evaluation of indentation techniques for measuring fracture toughness: I, direct crack measurements. *J. Am. Ceram. Soc.* **1981**, *64*, 533–538. [CrossRef]
26. Koumoulos, E.P.; Tofail, S.A.M.; Silien, C.; De Felicis, D.; Moscatelli, R.; Dragatogiannis, D.A.; Bemporad, E.; Sebastiani, M.; Charitidis, C.A. Metrology and nano-mechanical tests for nano-manufacturing and nano-bio interface: Challenges & future perspectives. *Mater. Des.* **2018**, *137*, 446–462.
27. Bull, S.J. Nanoindentation of coatings. *J. Phys. D Appl. Phys.* **2005**, *38*, R393. [CrossRef]
28. Iriarte, G.F.; Engelmark, F.; Ottosson, M.; Katardjiev, I. Influence of deposition parameters on the stress of magnetron sputter-deposited AlN thin films on Si (100) substrates. *J. Mater. Res.* **2003**, *18*, 423–432. [CrossRef]
29. Jones, A.M.; McCabe, A.R.; Bull, S.J.; Dearnaley, G. The effects of deposition temperature and interlayer thickness on the adhesion of ion-assisted titanium nitride films produced with yttrium metal interlayers. *Nucl. Instrum. Methods Phys. Res. Sect. B Beam Interact. Mater. Atoms* **1993**, *80–81*, 1397–1401. [CrossRef]
30. Gerth, J.; Wiklund, U. The influence of metallic interlayers on the adhesion of PVD TiN coatings on high-speed steel. *Wear* **2008**, *264*, 885–892. [CrossRef]
31. Suo, Z.; Hutchinson, J.W. Interface crack between two elastic layers. *Int. J. Fract.* **1990**, *43*, 1–18. [CrossRef]

32. Ionescu, G.; Nae, I.; Ripeanu, R.; Dinita, A.; Stan, G. Studies on Tribological Behavior of Aluminum Nitride-Coated Steel. In *IOP Conference Series: Materials Science and Engineering*; IOP Publishing: Bristol, UK, 2017; p. 012052.
33. Steinwandel, J.; Hoeschele, J.; Staneff, T.; Heuberger, A. Method and Apparatus for Aluminum Nitride Coating of a Contact Surface, Especially a Cylinder Contact Surface of a Crankcase Made of an Aluminum Basic Alloy. U.S. Patents 6,180,189, 30 January 2001.
34. Pina Bhatt, D.H. Performance Evaluation of TiAlN and CrN Coated Silicon Wafer by Magnetron Sputtering Method. *Int. J. Nanotechnol. Appl.* **2017**, *11*, 167–177.

nanomaterials

MDPI

Article

Investigation of the Surface of Ga–Sn–Zn Eutectic Alloy by the Characterisation of Oxide Nanofilms Obtained by the Touch-Printing Method

Alexandra Dobosz [1,*], Torben Daeneke [2], Ali Zavabeti [2], Bao Yue Zhang [2], Rebecca Orrell-Trigg [2], Kourosh Kalantar-Zadeh [2], Anna Wójcik [1], Wojciech Maziarz [1] and Tomasz Gancarz [1]

1 Institute of Metallurgy and Materials Science, Polish Academy of Sciences, 30-059 Krakow, Poland; a.wojcik@imim.pl (A.W.); w.maziarz@imim.pl (W.M.); tomasz.gancarz@imim.pl (T.G.)

2 School of Engineering, RMIT University, Melbourne, Victoria 3001, Australia; torben.daeneke@rmit.edu.au (T.D.); ali.zavabeti@rmit.edu.au (A.Z.); baoyue.zhang@student.rmit.edu.au (B.Y.Z.); s3486475@student.rmit.edu.au (R.O.-T.); k.kalantar-zadeh@unsw.edu.au (K.K.-Z.)

* Correspondence: a.dobosz@imim.pl; Tel.: +48-122952815

Received: 31 December 2018; Accepted: 6 February 2019; Published: 9 February 2019

Abstract: Ga–Sn–Zn eutectic alloy is a non-toxic liquid metal alloy which could be used in a multitude of applications, including as a heat transfer agent, in gas sensing, and in medicine. Alloys containing gallium readily oxidise in air, forming a thin oxide layer that influences the properties of liquid metals and which has not been studied. In this study, the oxide layer formed on Ga–Sn–Zn alloy was transferred at room temperature onto three substrates—quartz, glass and silicon. The contact angle between the liquid alloy and different substrates was determined. The obtained thin oxide films were characterised using atomic force microscopy, X-ray photon spectroscopy, and optical and transmission electron microscopy. The contact angle does not influence the deposition of the layers. It was determined that it is possible to obtain nanometric oxide layers of a few micrometres in size. The chemical composition was determined by XPS and EDS independently, and showed that the oxide layer contains about 90 atom % of gallium with some additions of tin and zinc. The oxides obtained from the eutectic Ga–Sn–Zn liquid alloys appear to be nanocrystalline.

Keywords: liquid alloys; 2D materials; thin films; Ga–Sn–Zn alloys; gallium alloys; nanoanalysis

1. Introduction

Liquid metal alloys based on gallium are a group of metallic alloys, liquid at room temperature, that could possibly replace toxic materials such as mercury or alloys based on lead. The most commonly used alloy based on gallium is the eutectic Ga–In–Sn alloy Galinstan, which has potential to be used in many fields including electrical engineering, computer engineering [1] and medicine [2,3]. Eutectic Ga–In–Sn alloy is currently being studied intensively as an effective micro-cooling agent with superior power-handling properties [4–6]. Other alloys have been proposed as a cheaper alternative to Galinstan, based on gallium, tin and zinc. These include Ga–Sn–Zn eutectic alloy [7]. Liquid metal alloys have a range of advantages and could find uses in many different applications. Those alloys exhibit excellent thermal and electrical conductivity, are safe to handle due to almost no vapour pressure [8], and exhibit a high supercooling effect [7]. Moreover, the surface of liquid gallium and its alloys oxidises easily, forming an oxide skin on the alloy [9–11]. The oxide skin on pure gallium has been investigated [12,13], but there are only limited data on oxides forming on alloys containing gallium as the main component. In the case of pure gallium, according to reference [12], the thickness of the solid oxide film is about 0.5 nm. For ambient oxygen pressure and temperatures, it was proposed that isotropic diffusion aggregation was the process by which the gallium oxide grows and expands on the surface of liquid gallium [13]. The oxide first grows in a fractal-like pattern, then interfacial smoothening occurs, which may be caused by insoluble impurities in the oxide [13].

This study focuses on gallium-based alloys with tin and zinc additions. The Ga–Sn–Zn ternary system is formed from three simple eutectic systems [14–16]. The Ga–Sn–Zn eutectic alloy, with a composition of 90.15 of Ga, 6.64 of Sn, and 3.21 of Zn (atom %) corresponding to 86.3, 10.8, and 2.9 (wt %), has a melting point of 288 K and a solidification point of 266.5 K [7]. It has been found that adding a fourth element to the ternary eutectic system can increase the difference between the melting and solidification temperature [17,18]. By adding 0.5 atom % aluminum to Ga–Sn–Zn, the difference between the two temperatures increases from 20.6 K to 65.8 K, with a melting point of 301.3 K and a solidification point of 235.5 [17]. Similarly, indium additions cause the solidification temperature to drop even further to 231 K in the case of 14.7 atom % In, with a low melting point of 289 K [18].

As gallium alloys remain liquid over a large temperature range, touch-printing can be used in order to obtain nanometric films from the surface of the alloys [19]. The method developed in reference [19] allows the transfer of oxidised material on the liquid metal surface to virtually any substrate. The method has been used to obtain 2D Ga_2O_3, HfO_2, Gd_2O_3, and Al_2O_3 [19]. Due to the low costs of fabrication of these nanometric films, the materials could be used in many important fields of science and industry, including as ultra-thin insulator dielectrics in field-effect transistors, for energy storage, and in gas sensing [20]. Using gallium alloys with additions of different elements could lower the synthesis temperature of chosen oxides; however, it is crucial to understand how those additions will affect the final composition of the oxide layer and which elements will be incorporated into the obtained material. In this work, we aimed to obtain 2D layers using the touch-printing technique on silicon, glass and quartz, in order to characterise the oxide layer formed on the surface of liquid eutectic Ga–Sn–Zn alloy using atomic force microscopy, X-ray photon spectroscopy, and optical and transmission electron microscopy.

2. Experimental

The Ga–Sn–Zn eutectic alloy was prepared in IMMS PAS using pure Ga (99.99999% PPM Pure Metals), Sn and Zn (99.999% Alfa Aeaser) in a glovebox under a protective atmosphere of high-purity argon (99.9999% Air Products). The oxygen, nitrogen and water vapour concentrations were lower than 0.1 ppm in order to avoid the oxidation of the prepared liquid alloys. The elements were melted together and homogenised in a graphite crucible in an electrical furnace placed in the glovebox.

The thin films were obtained at room temperature in ambient air by placing a drop of liquid metal on a glass substrate. In the next step, the drop was touched with the chosen substrate. The oxide skin attached itself onto the substrate with some residual liquid metal. The liquid metal was then cleaned off the substrate by dipping the material in boiling water and wiping it with Kimwipes. The material thus cleaned and dried was analysed. Various substrates were tested, including glass, quartz and silicon. The substrates were cleaned twice before the process, in isopropanol, ethanol and water.

In order to determine the substrate on which the deposition would be easiest, and, consequently, on which the best quality of films could be obtained, the wettability of the Ga–Sn–Zn eutectic alloy on different substrates was measured at room temperature in ambient air, outside the glovebox, in order to imitate the same conditions in which the layers are deposited on the substrates. The contact angles were measured using the static sessile drop method in IMMS PAS.

Moreover, the obtained thin films were studied at RMIT University by atomic force microscopy (Dimension Icon-Bruker with scan-assist software), and the thickness of the layers was determined. Due to the thickness of the layers, X-ray photoelectron spectroscopy (Thermo Scientific K Alpha XPS) was used to analyse the chemical composition, as this is a suitable method in the case of nanometric materials. The X-ray source was a monochromated Al Kα source. The samples on 3 mm TEM copper grids were also studied by transmission electron microscopy using a Tecnai G2 F20 operating at 200 keV in IMMS PAS. The chemical composition was also assessed using an energy-dispersive X-ray (EDX) microanalyser in TEM.

3. Results and discussion

The contact angle between the Ga–Sn–Zn eutectic and the substrates, namely, silicon, glass and quartz, was analysed first. For each material, the contact angles were measured three times. Representative images of the contact angles are presented in Figure 1.

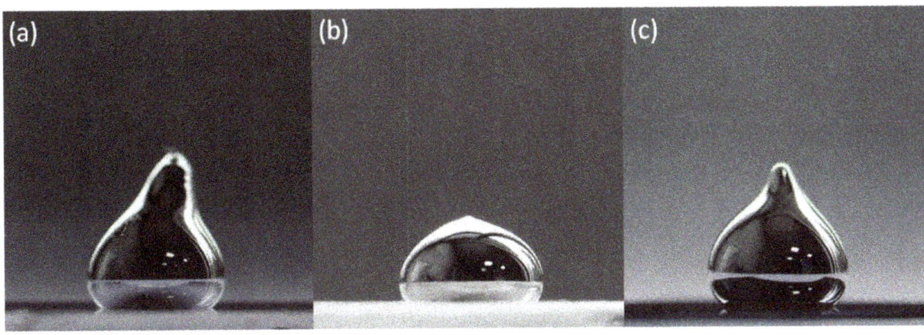

Figure 1. Representative contact angles for (**a**) glass, (**b**) quartz and (**c**) silicon.

The obtained contact angles were as follows: $132 \pm 1°$ for glass, $118 \pm 2°$ for quartz and $134 \pm 1°$ in the case of silicon. From the conducted measurements, it can be concluded that the wetting behaviour of Ga–Sn–Zn eutectic alloy is similar in the cases of glass and silicon. The obtained values in the case of silicon are similar to the value obtained in reference [21] for Galinstan on silicon (139° at 303 K). However, it was found in reference [22] that Galinstan was found to be non-wetting on different substrates, including glass, when not oxidised. Furthermore, it was determined that the moment the oxide skin forms on the surface of Galinstan, the alloy sticks on surfaces and gives the illusion that Galinstan wets the substrates [22].

In relation to 2D materials, our investigations are focused on obtained layers which were analysed using optical microscopy, scanning electron microscopy, atomic force microscopy and X-ray photoelectron spectroscopy. An optical micrograph of the nanolayers obtained on silicon is presented in Figure 2.

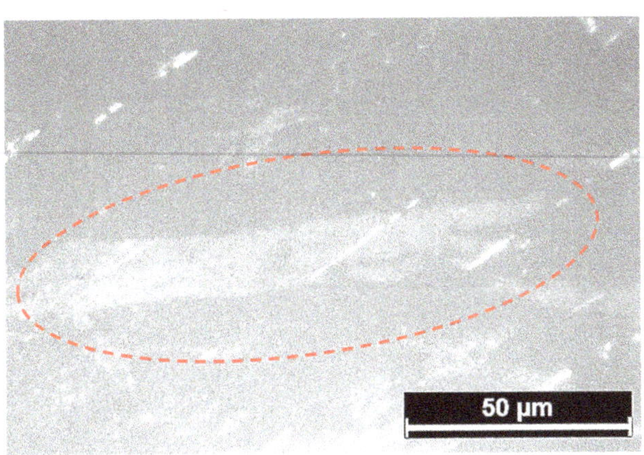

Figure 2. Optical microphotograph of thin film obtained from Ga–Sn–Zn eutectic alloy on glass (the layer has been circled).

As can be seen from Figure 2, the oxide layer obtained by the touch-printing method is continuous, although there are some traces of metal left on the surface of the sample. The same microstructure is observed in the cases of all different substrates. In the next step, the layers were further analysed using atomic force microscopy. For example, the layers obtained from the Ga–Sn–Zn eutectic alloy on the silicon substrate are presented in Figure 3.

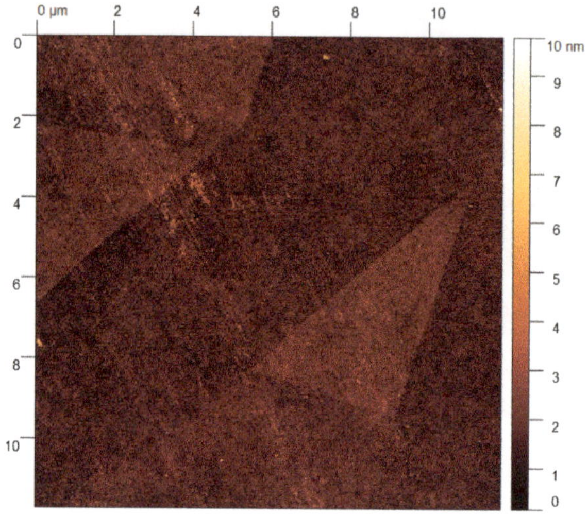

Figure 3. Image from AFM of nanolayers obtained from Ga–Sn–Zn eutectic alloy on silicon.

Using AFM, the layer appears to be continuous. The measured thickness of the layers obtained from Ga–Sn–Zn eutectic alloy is about 1.5 nm. Due to the low thickness of the sample, comparable to background noise, the line depth profile is not shown. The layers obtained on glass and quartz were of similar thicknesses, ranging from 1.5 to 3 nm in the thinnest parts. The thickness of the thin films is 3 times the thickness of the layer of the oxide skin on pure liquid gallium reported in reference [12].

The chemical composition was analysed using X-ray photoelectron spectroscopy. The XPS spectra are presented in Figure 4. In the case of the gallium, the peak at 20.2 eV corresponds to Ga_2O_3, while the peak at around 25 eV is the O2s peak, which is in the same range as Ga. The peak observed at 18.2 eV is elemental Ga, which may indicate some inclusions of gallium in the obtained material. In the case of the zinc spectrum, the analysis is more complicated due to the fact that the difference between the binding energies of elemental zinc and zinc oxide is less than 1 eV. However, the main zinc peak is located at 1022 eV, which is closer to the ZnO oxide binding energy. In the case of tin, the peak around 485 eV can be identified as corresponding to metallic tin, while the peak at 486.7 eV corresponds to tin oxide SnO_2.

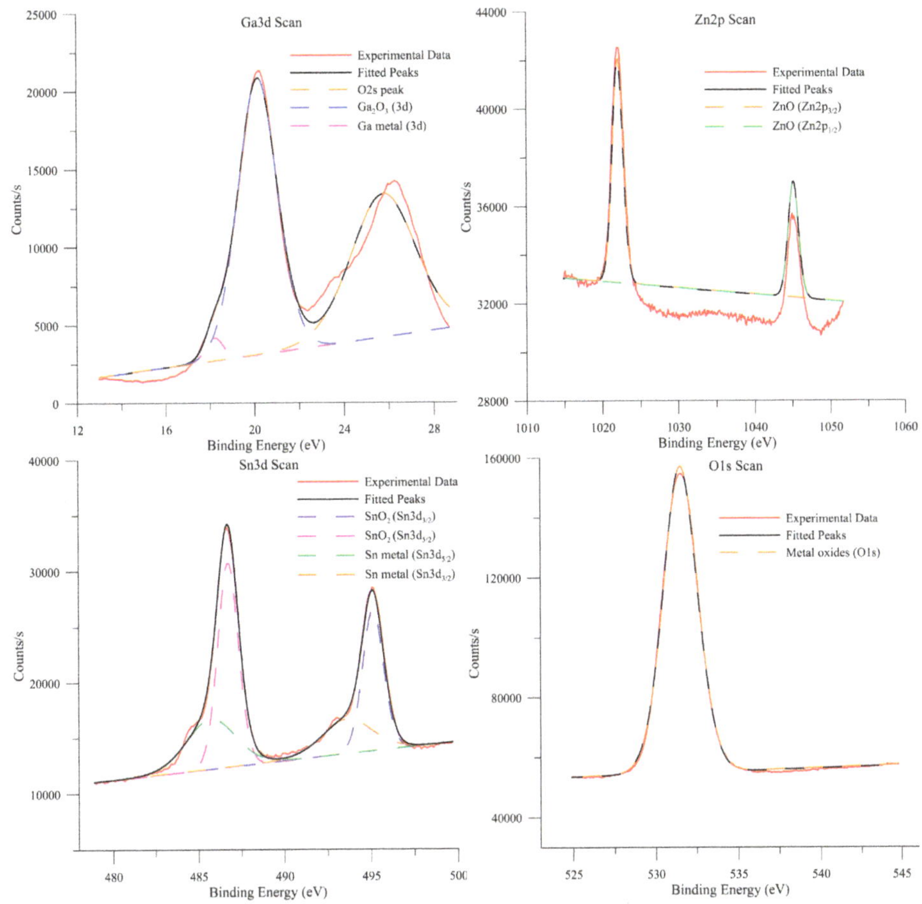

Figure 4. XPS spectra of Ga, Zn, Sn and O in oxide layers obtained from Ga–Sn–Zn eutectic alloy.

It was determined that the chemical composition of the touch-printed layers is 88.1 atom % of gallium, 6.9 atom % of tin and 5.0 atom % of zinc. When oxygen is taken into account, the composition of the layer is as follows: 19.7 atom % of gallium, 1.6 atom % of tin, 1.1 atom % of zinc and 77.6 atom % of oxygen. It is crucial to note that, according to simple thermodynamics, the Ga_2O_3 ($\Delta G_f = -998.3$ kJ/mol) phase is more likely to form than SnO_2 or ZnO ($\Delta G_f = -515.8$ and -350.5 kJ/mol, respectively) [23]. However, the presented results indicate the presence of not only gallium oxide but also tin and zinc oxides. It is possible that the tin and zinc substitute for gallium atoms in the Ga_2O_3 oxide lattice, or that some form of segregation occurs on the surface of the liquid alloy. It should also be taken into account that some tin and zinc oxide could be formed after the transfer occurs, and the metal left on the substrate before cleaning oxides.

The TEM images are shown in Figure 5. The presented image shows a piece of the layer of approximately 1.2 × 0.8 μm. The oxide appears to be nanocrystalline. The material obtained on the TEM grid is much smaller than the material that can be obtained on non-porous surfaces, i.e., glass, silicon or quartz. Even though TEM grids with carbon film were used, the material does not transfer well onto the grids and does not appear in the form of clean sheets, but rather as folded pieces of oxide skin. EDS analysis was conducted, and the results are shown in Figure 6.

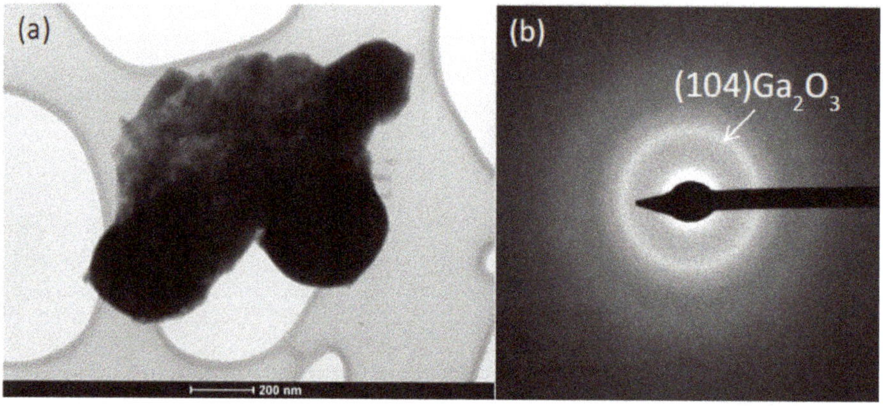

Figure 5. (**a**) TEM bright field image and (**b**) the corresponding selected area diffraction pattern.

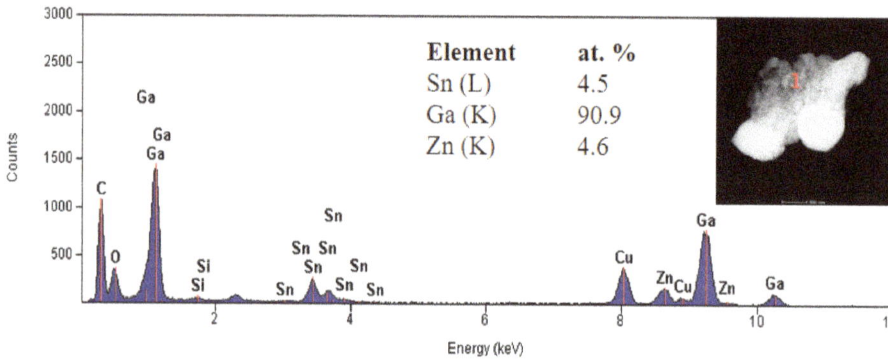

Figure 6. EDS point analysis of the chemical composition of the material. Insert: dark field image.

Although the XPS and EDS results of the chemical composition investigation vary slightly, it can be concluded that they are consistent and indicate that the oxide layer is formed mainly of gallium oxide.

4. Conclusions

It is possible to obtain materials by transferring the oxide formed on liquid Ga–Sn–Zn alloy at room temperature onto different substrates, namely, quartz, glass and silicon. The contact angle between the liquid metal and the studied substrate (being in the range between 117° and 134°) does not affect the conducted measurements, although in this respect it could be interesting to study more hydrophobic and more hydrophilic surfaces. It is difficult to obtain macroscopic clear and smooth layers, due to folding of the oxide layer and oxidation of the metal on the analysed substrate. However, as can be seen from the atomic force image, it is possible to obtain smooth nanometric layers of a few micrometres in length. The XPS and EDS results suggest that the thin films obtained from eutectic Ga–Sn–Zn alloy contain not only gallium oxide, which should form according to thermodynamics, but also smaller quantities of tin and zinc oxides. The selective area diffraction pattern shows the presence of Ga_2O_3. It has not been determined whether the tin and zinc are substituting for gallium in the oxide or forming small patches of tin and zinc oxide in the main phase of gallium oxide. The study of liquid metal alloy surfaces should be undertaken as an important step in future research concerning alloys with low melting points, as it may prove crucial in many studied applications.

Author Contributions: Conceptualisation, K.K.-Z. and T.G.; methodology, W.M.; validation, W.M.; investigation, A.D., A.W., A.Z. and B.Y.Z.; resources, T.G. and K.K.-Z.; writing—original draft preparation, A.D.; writing—review and editing, A.D., T.G. and B.Y.Z.; visualisation, A.D.; supervision, T.G. and T.D.; project administration, T.D.; funding acquisition, A.D. and T.G.

Funding: This work was financed by the National Science Centre Poland grant 2016/21/B/ST8/00324 "Design and physicochemical, thermal properties of low temperature metal alloys based on gallium" in 2017-2019. The research was co-financed by the European Union from resources of the European Social Fund (Project No.WND-POWR.03.02.00-00-I043/16.

Conflicts of Interest: The authors declare no conflict of interest.

References

1. Lorenzin, N.; Abanades, A. A review on the application of liquid metals as heat transfer fluid in Concentrated Solar Power technologies. *Int. J. Hydrog. Energy* **2016**, *41*, 6990–6995. [CrossRef]
2. Yue, L.; Hu, Q.; Lin, Y.; Pacardo, D.B.; Wang, C.; Sun, W.; Ligler, F.S.; Dickey, M.D.; Gu, Z. Transformable liquid-metal nanomedicine. *Nat. Commun.* **2015**, *6*, 10066.
3. Kyle, A.M.; Leahy-Glass, N.; Combs, W.; March, K.L. Potential of gallium-based leads for cardiac rhythm management devices. In Proceedings of the 2011 Annual International Conference of the IEEE Engineering in Medicine and Biology Society, Boston, MA, USA, 30 August–3 September 2011; pp. 341–344.
4. Irshad, W.; Peroulis, D. A silicon-based galinstan magnetohydrodynamic pump. In Proceedings of the 9th International Workshop on Micro and Nanotechnology for Power Generation and Energy Conversion Applications, Washington DC, USA, 1–4 December 2009; pp. 127–129.
5. Hodes, M.; Zhang, R.; Lam, L.S.; Wilcoxon, R.; Lower, N. On the potential of galinstan-based minichannel and minigap cooling. *IEEE Trans. Compon. Packag. Manuf. Technol.* **2014**, *4*, 46–56. [CrossRef]
6. Zhu, J.Y.; Tang, S.Y.; Khoshmanesh, K.; Ghorbani, K. An integrated liquid cooling system based on galinstan liquid metal droplets. *ACS Appl. Mater. Interfaces* **2016**, *8*, 2173–2180. [CrossRef] [PubMed]
7. Dobosz, A.; Plevachuk, Y.; Sklyarchuk, V.; Sokoliuk, B.; Gancarz, T. Thermophysical properties of the liquid Ga–Sn–Zn eutectic alloy. *Fluid Phase Equilib.* **2016**, *465*, 1–9. [CrossRef]
8. Fowler, B.A.; Sexton, M.J. *Handbook on the Toxicology of Metals*, 4th ed.; Nordberg, G.F., Fowler, B.A., Nordberg, M., Eds.; Academic Press: Cambridge, MA, USA, 2014; pp. 787–795.
9. Ladd, C.; So, J.H.; Muth, J.; Dickey, M.D. 3D printing of free-standing liquid metal microstructures. *Adv. Mater.* **2013**, *25*, 5081–5085. [CrossRef] [PubMed]
10. Boley, J.W.; White, E.L.; Chiu, G.T.C.; Kramer, R.K. Direct writing of gallium-indium alloy for stretchable electronics. *Adv. Funct. Mater.* **2014**, *24*, 3501–3507. [CrossRef]
11. Dickey, M.D. Emerging applications of liquid metals featuring surface oxides. *ACS Appl. Mater. Interfaces* **2014**, *6*, 18369–18379. [CrossRef] [PubMed]
12. Regan, M.J.; Tostmann, H.; Pershan, P.S.; Magnussen, O.M.; DiMasi, E.; Ocko, B.M.; Deutsch, M. X-ray study of the oxidation of liquid-gallium surfaces. *Phys. Rev. B* **1997**, *55*, 10786–10790. [CrossRef]
13. Chabala, J.M. Oxide-growth kinetics and fractal-like patterning across liquid gallium surfaces. *Phys. Rev. B* **1992**, *46*, 11346–11357. [CrossRef]
14. Anderson, T.J.; Ansara, I. The Ga-Sn (gallium-tin) system. *J. Fluid Phase Equilib.* **1992**, *13*, 181–189. [CrossRef]
15. Dutkiewicz, J.; Moser, Z.; Zabdyr, L.; Gohil, D.D.; Chart, T.G.; Ansara, I.; Girard, C. The Ga-Zn (gallium-zinc) system. *Bull. Alloy Phase Diagr.* **1990**, *11*, 77–82. [CrossRef]
16. Moser, Z.; Dutkiewicz, J.; Gąsior, W.; Salawa, J. The Sn– Zn (Tin-Zinc) system. *Bull. Alloy Phase Diagr.* **1985**, *6*, 330–334. [CrossRef]
17. Dobosz, A.; Plevachuk, Y.; Sklyarchuk, V.; Sokoliuk, B.; Gancarz, T. The application of liquid metals in cooling systems: A study of the thermophysical properties of eutectic Ga-Sn-Zn with Al additions. *Int. J. Heat Mass Transf.* **2018**, *126*, 414–420. [CrossRef]
18. Dobosz, A.; Plevachuk, Y.; Sklyarchuk, V.; Sokoliuk, B.; Tkach, O.; Gancarz, T. The thermophysical properties of eutectic Ga-Sn-Zn with In additions. *J. Mol. Liq.* **2018**, *271*, 942–948. [CrossRef]
19. Zavabeti, A.; Ou, J.Z.; Carey, B.J.; Syed, N.; Orrell-Trigg, R.; Mayes, E.L.; Xu, C.; Kavehei, O.; O'mullane, A.P.; Kaner, R.B.; Kalantar-Zadeh, K. A liquid metal reaction environment for the room-temperature synthesis of atomically thin metal oxides. *Science* **2017**, *358*, 332–335. [CrossRef] [PubMed]

20. Sberveglieri, G. Recent developments in semiconducting thin-film gas sensors. *Sens. Actuators B Chem.* **1995**, *23*, 103–109. [CrossRef]
21. Davis, E.; Ndao, S. On the Wetting States of Low Melting Point Metal Galinstan® on Silicon Microstructured Surfaces. *Adv. Eng. Mater.* **2018**, *20*, 1700829. [CrossRef]
22. Liu, T.; Sen, P.; Kim, C.J. Characterization of nontoxic liquid-metal alloy galinstan for applications in microdevices. *J. Microelectromech. Syst.* **2012**, *21*, 443–450. [CrossRef]
23. Dean, J.A. *Lange's handbook of chemistry*, 15th ed.; McGraw-Hill, Inc.: New York, NY, USA, 1999; pp. 6.1–6.158.

nanomaterials

MDPI

Article

Tailoring Upconversion and Morphology of Yb/Eu Doped Y_2O_3 Nanostructures by Acid Composition Mediation

Daniela Nunes [1,*], Ana Pimentel [1,*], Mariana Matias [1], Tomás Freire [1], A. Araújo [2], Filipe Silva [2], Patrícia Gaspar [2], Silvia Garcia [2], Patrícia A. Carvalho [3], Elvira Fortunato [1] and Rodrigo Martins [1,*]

1. i3N/CENIMAT, Department of Materials Science, Faculty of Sciences and Technology, Universidade NOVA de Lisboa and CEMOP/UNINOVA, Campus de Caparica, 2829-516 Caparica, Portugal; ms.matias@campus.fct.unl.pt (M.M.); tm.freire@campus.fct.unl.pt (T.F.); emf@fct.unl.pt (E.F.)
2. INCM, Imprensa Nacional-Casa da Moeda, Av. António José de Almeida, 1000-042 Lisboa, Portugal; Andreia.Cardoso@incm.pt (A.A.); filipe.silva@incm.pt (F.S.); patricia.gaspar@incm.pt (P.G.) Silvia.Garcia@incm.pt (S.G.)
3. SINTEF Materials Physics, Forskningvein 1, 0373 Oslo, Norway; patricia.carvalho@sintef.no

* Correspondence: daniela.gomes@fct.unl.pt (D.N.); acgp@campus.fct.unl.pt (A.P.); rm@uninova.pt (R.M.); Tel.: +351-21-2948562 (D.N. & A.P. & R.M.); Fax: +351-21-2948558 (D.N. & A.P. & R.M.)

Received: 29 December 2018; Accepted: 4 February 2019; Published: 9 February 2019

Abstract: The present study reports the production of upconverter nanostructures composed by a yttrium oxide host matrix co-doped with ytterbium and europium, i.e., Y_2O_3:Yb^{3+}/Eu^{3+}. These nanostructures were formed through the dissociation of yttrium, ytterbium and europium oxides using acetic, hydrochloric and nitric acids, followed by a fast hydrothermal method assisted by microwave irradiation and subsequent calcination process. Structural characterization has been carried out by X-ray diffraction (XRD), scanning transmission electron microscopy (STEM) and scanning electron microscopy (SEM) both coupled with energy dispersive X-ray spectroscopy (EDS). The acid used for dissociation of the primary oxides played a crucial role on the morphology of the nanostructures. The acetic-based nanostructures resulted in nanosheets in the micrometer range, with thickness of around 50 nm, while hydrochloric and nitric resulted in sphere-shaped nanostructures. The produced nanostructures revealed a homogeneous distribution of the doping elements. The thermal behaviour of the materials has been investigated with in situ X-Ray diffraction and differential scanning calorimetry (DSC) experiments. Moreover, the optical band gaps of all materials were determined from diffuse reflectance spectroscopy, and their photoluminescence behaviour has been accessed showing significant differences depending on the acid used, which can directly influence their upconversion performance.

Keywords: microwave synthesis; oxide dissociation; doping; rare earth ions; upconversion

1. Introduction

Upconverter materials are characterized by having the ability to emit photons with higher energy than the photons that were absorbed [1]. The most studied upconverters use rare earth doped luminescent materials. Rare earth elements belong to the lanthanides group in the periodic table and typically are trivalent ions (Ln^{3+}) in which the 4f inner shell is filled up to 14 electrons, presenting a $4f^n5s^25p^6$ electron configuration (with n = 0–14) [1]. The partial filled 4f shell is the responsible for the optical and magnetic properties presented by this type of materials, and for the generation of radiation of different wavelengths, depending on the dopants used [2]. The most common used lanthanides elements are: Eu^{2+} and Ce^{3+} (broad band emitting due to the 5d $\rightarrow$ 4f transition) and Eu^{3+},

Tb^{3+}, Gd^{3+}, Yb^{3+}, Dy^{3+}, Sm^{3+}, Tm^{3+}, Er^{3+}, and Nd^{3+} (narrow band emitting due to transitions between 4f levels) [3]. Most of them present visible emissions, and some are described in Table 1 [4–6].

Table 1. Transitions of lanthanide elements [3–6].

Element	Color of the Visible Luminescence	Transition	Intensity
Ce, cerium	UV	5d → $^2F_{5/2}$ (300–450 nm)	n.a.
Pr, praseodymium	Red	$^1D_2 \rightarrow {}^4H_4$ (600 nm) $^3P_0 \rightarrow {}^3F_2$ (700 nm)	Weak
Nd, neodymium	Infra-red	$^4F_{3/2} \rightarrow {}^4I_{9/2}$ (900 nm)	n.a.
Sm, samarium	Orange–red	$^4G_{5/2} \rightarrow {}^6H_{7/2}$ (600 nm)	Medium
Eu, europium	Red–orange	$^5D_0 \rightarrow {}^7F_4$ (720 nm) $^5D_0 \rightarrow {}^7F_3$ (650 nm) $^5D_0 \rightarrow {}^7F_2$ (615 nm) $^5D_0 \rightarrow {}^7F_1$ (590 nm) $^5D_0 \rightarrow {}^7F_0$ (580 nm)	Strong
Gd, gadolinium	UV	$^6P_{7/2} \rightarrow {}^8S_{7/2}$ (310 nm)	Strong
Td, terbium	Green–orange	$^5D_4 \rightarrow {}^7F_5$ (540 nm) $^5D_4 \rightarrow {}^7F_4$ (580 nm) $^5D_4 \rightarrow {}^7F_3$ (620 nm) $^5D_4 \rightarrow {}^7F_2$ (650 nm) $^5D_4 \rightarrow {}^7F_1$ (660 nm)	Strong
Dy, dysprosium	Yellow	$^4F_{9/2} \rightarrow {}^6H_{13/2}$ (570 nm) $^4I_{15/2} \rightarrow {}^6H_{13/2}$ (540 nm)	Medium
Ho, holmium	Red	$^5F_5 \rightarrow {}^5I_8$ (650 nm)	Weak
Er, erbium	Green	$^4S_{3/2} \rightarrow {}^4I_{15/2}$ (545 nm) $^4F_{9/2} \rightarrow {}^4I_{15/2}$ (660 nm)	Weak
Tm, thulium	Blue	$^1G_4 \rightarrow {}^3H_6$ (470 nm)	Weak
Yb, ytterbium	Infra-red	$^2F_{5/2} \rightarrow {}^7F_{7/2}$ (980 nm)	n.a.

n.a.—not available.

In the past few years, these rare earth oxides have attracted an increasing attention due to their possible use in several applications such as luminescence devices [7,8], color displays [9,10], optical detectors [11,12], telecommunications [5], upconverter lasers and photonics [13–15], catalysis [16], biomedicine [17,18] and solar cells [1,19,20], mainly due to their chemical, electronic, and optical properties resulting from the 4f electronic shells [21]. These properties are greatly dependent on the crystal structure, morphology and also on the chemical composition (type of dopants used), which are very sensitive to the bonding states of rare earth ions [21].

Yttrium oxide, Y_2O_3, is a rare earth oxide with a relatively high band gap value, around 5.8 eV [22], low phonon frequency and ionic radii [2]. Y_2O_3 is considered to be one of the greatest luminescent oxide phosphor materials available, mainly due to its high luminescence efficiency, chemical, thermal and photochemical stabilities but also due to its high color purity [23–25], presenting a large dielectric constant and good infrared transmission [22].

Y_2O_3 is a host material (matrix) and its wide band gap makes it attractive for several optical applications in the visible and UV ranges. Doping Y_2O_3 with other rare earth elements can bring some advantages, since they can emit within its optical window and not suffer quenching effects [26].

Ytterbium, Yb^{3+} is often used as a sensitizer due to its higher absorption cross-section that corresponds to 980 nm excitation, which will excite other elements like europium, Eu^{3+} [2]. Europium, on the other hand, is a trivalent material that contains ions from the rare earth and from the transition-metal series [27]. Doping Y_2O_3 with Eu^{3+} is considered to be one of the best red oxide phosphors due to their chemical stabilities and luminescence efficiencies [21,23]. The visible emission spectrum of Eu^{3+} is composed by $^5D_0 \rightarrow {}^7F_2$, $^5D_0 \rightarrow {}^7F_1$ and $^5D_0 \rightarrow {}^7F_0$ [4]. The ratio between transitions $^5D_0 \rightarrow {}^7F_2$ and $^5D_0 \rightarrow {}^7F_1$ intensities, is an indicative of the resulting emission color. The higher the ratio, the more intense into the orange color will be the luminescence [4].

Y_2O_3 nanoparticles doped with Yb^{3+} and Eu^{3+} ions are expected to possess the ability to absorb near-infrared radiations (NIR) and upconvert it into visible radiation [28]. Yb^{3+} will work as a sensitizer (possessing one energy level around 980 nm, with a lifetime of ~2 ms, ideal for infrared to visible upconvertion), that will transfer the NIR energy to the activator Eu^{3+}, with an emission wavelength shorter than the NIR excitation [1,28].

In terms of nanostructure shape, the synthesis of nanosized spherical phosphors particles brings the advantage of enhanced brightness and high resolution in upconversion applications [29], but also high packing densities and low light scattering. Therefore, it is suggested that the ideal morphology for a phosphor nanoparticle is a perfect spherical shape, with a narrow size distribution [29]. Several methods have already been reported for the synthesis of phosphors Y_2O_3 nanostructures aiming to obtain spherical shaped particles with homogeneous sizes and distributions, such as spray pyrolysis [23], sol-gel [30,31], hydrothermal/solvothermal methods [29], solution combustion [32,33], precipitation [34–36] and template method. Microwave assisted synthesis has also been reported for the production of Y_2O_3 materials [24,37,38], having numerous advantages, such as its celerity, reproducibility and cost-efficiency, in such a way that in the last few years it has been widely used on the synthesis of this oxide and several other types of oxides [39–41].

The present work focuses on the dissociation of primary oxides using different acids and further microwave synthesis of nanostructures based on Y_2O_3 doped with Yb and Eu. This synthesis route required minimal time to effectively dope the nanostructured matrix (15 min), which makes this a fast approach to produce such materials. The obtained nanostructures were fully characterized by X-ray diffraction (XRD), scanning transmission electron microscopy (STEM) and scanning electron microscopy (SEM), differential scanning calorimetry (DSC), Ultra Violet- Visible – Near Infrared (UV-VIS-NIR) spectroscopy, and photoluminescence.

2. Synthesis and Characterization of Yb/Eu Doped Y_2O_3 Nanostructures

2.1. Materials

The influence of different acids on the production of upconverter nanostructures under microwave irradiation was studied. For all the materials produced, the microwave synthesis was repeated at least 3 times for each set of samples produced, showing that this technique is highly reproducible, and presenting, within each set, the same type of structure, morphology and composition.

Regardless of the chosen acid, the basis of this synthesis is composed by a host matrix of yttrium oxide (Y_2O_3, CAS: 1314-36-9, purity 99.99%, with a particle size < 50 nm, from Sigma Aldrich, (Sigma-Aldrich Chemie Gmbh, Munich, Germany), and the dopants coming from the ytterbium oxide (Yb_2O_3, CAS: 1314-37-0, purity 99.7%, with a particle size < 100 nm, from Sigma Aldrich) and europium oxide (Eu_2O_3, CAS: 1308-96-9, purity 99.5%, with a particle size < 150 nm, from Sigma Aldrich), here named as the primary oxides. Several different acids have been used to dissolve the primary oxides and convert them to the form of nitrates: acetic acid (CH_3COOH, CAS: 64-19-7, from Fisher Chemical (Fisher Scientific GmbH, Schwerte, Germany), hydrochloric acid (HCl, CAS: 7647-01-0, 37%, from Fisher Chemical) and nitric acid (HNO_3 CAS: 7697-37-2, 69%, from Labkem(Blanc-Labo SA, Lonay, Switzerland). All the initial chemicals were used without further purification.

2.2. Yb/Eu Doped Y_2O_3 Synthesis

Three solutions were prepared, each of them with a different acid, *i.e.*, acetic, hydrochloric and nitric acids. In this way, it became possible to study the influence of different acids compositions in the final materials' outcomes, resorting to a hydrothermal synthesis assisted by microwave irradiation.

In a typical synthesis, 0.05 M of Y_2O_3, 0.01 M of YbO_3 and 0.0075 M of Eu_2O_3 were diluted in 5 mL of acid (CH_3COOH, HCl or HNO_3), resulting in the formation of a colorless solution. The solution was kept under stirring on a hot plate at 70 °C in ambient atmosphere, until complete evaporation. A white powder was then obtained.

Afterwards, 40 mL of H_2O and a most common used catalyst [2,42–45] was added to this powder and left stir for 10 minutes until complete dissociation. The solution was then placed in a Pyrex vessel and placed into a microwave heating system, Discover SP, from CEM (CEM Corporation, Matthews, NC, USA), fixing the maximum power up to 100 W with a temperature of 130 °C and 15 min synthesis time.

The obtained powders were then washed with deionized water and isopropyl alcohol. Between each washing process, the powder was centrifuged at 3000 rpm for 3 min. The annealing of the obtained nanostructures was carried out in a Nabertherm furnace (Nabertherm GmbH, Lilienthal, Germany), at a temperature of 700 °C for 6 h, under atmospheric conditions, with a heating ramp of 1 h. Figure 1 shows a schematic of Yb/Eu doped Y_2O_3 nanostructures production steps.

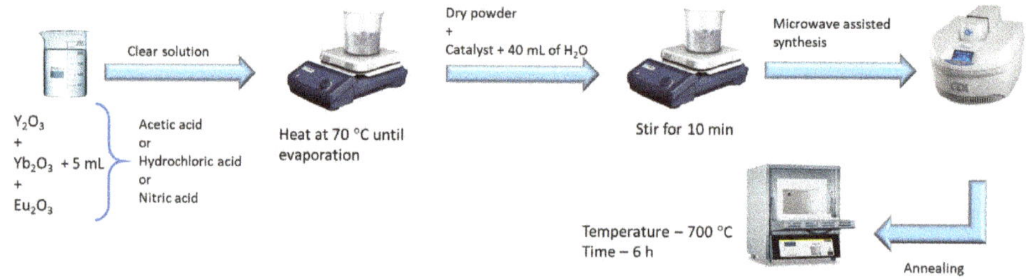

Figure 1. Schematic of Yb/Eu doped Y_2O_3 nanostructures produced by a hydrothermal process under microwave irradiation.

2.3. *Yb/Eu Doped Y_2O_3 Characterization*

All the characterization measurements were carried out in the powder form. The crystallinity and phase purity of as-synthesized and annealed materials were characterized by X-ray diffraction, using a PANalytical's X'Pert PRO MRD X-ray diffractometer (PANalytical B.V., Almero, The Netherlands), with a monochromatic CuKα, λ = 1.540598 Å, radiation source. The XRD measurements were carried out from 25° to 60°, with a scanning step size of 0.016°. The in situ diffractograms were collected in the same 2θ range and at temperatures in steps of 100 °C, from 300 °C to 1000 °C. The material was kept at each temperature step for at least half an hour to allow the stabilization and 5 consecutive scans were collected for inspection of structural modifications during this time. The temperature was increased with a rate of ~1.7 °C min^{-1}.

Differential Scanning Calorimetry measurements were carried out to the obtained powder, before annealing. It was used a simultaneous thermogravimetric analyzer, TGA-DSC-STA 449 F3 Jupiter, from Netzsch (Netzsch-Geratebau GmnH, Selb, Germany). Approximately 5 mg of powder was loaded into an open PtRh crucible and heated from room temperature to 1100 °C, with a heating rate of 10 °C min^{-1} under atmospheric conditions.

The morphology of all materials was analyzed by using a Scanning Electron Microscope (SEM) Carl Zeiss AURIGA CrossBeam Workstation instrument (Carl Zeiss Microscopy GmbH, Oberkochen, Germany), equipped with an Oxford energy dispersive X-ray spectrometer (EDS, Oxford Instruments Nanoanalysis, High Wycombe, UK).

Scanning Transmission Electron Microscopy (STEM) analyses of the nanostructures were carried out at 300 kV with a FEI Titan G2 60-300 instrument (Thermo Fisher Scientific, Hillsboro, OR, USA) equipped with a DCOR probe Cs-aberration corrector and a Super-X Bruker energy dispersive spectrometer with 4 silicon drift detectors. The size of 40 individual nanocrystals was determined through STEM images using ImageJ software (version 1.52k) [46].

For band gap determination, diffuse reflectance measurements were carried out at room temperature, using a Perkin Elmer lambda 950 UV/VIS/NIR spectrophotometer (Perkin Elmer, Inc., Waltham, MA, USA), with a diffuse light detection module (150 mm diameter integrating sphere, internally coated with Spectralon). The calibration of the system was performed by using a standard reflector sample (reflectance, R, of 1.00 from Spectralon disk). The reflectance (R) spectra was acquired in the range between 175 and 1200 nm.

Photoluminescence were also examined with a 976 nm laser (Avantes NIR light, Iso-Tech power source 1 W, 5 nm resolution) between 500 and 750 nm. All measurements were carried out under identical excitation/detection conditions, and for properly comparison of the PL intensities, the powder was weighted and pressed under the same pressure to obtain the same powder density.

3. Results and Discussion

The reactions between the different acids and the host matrix oxide (Y_2O_3) can be described by the following Equations (from 1 to 3) [47,48]:

$$\text{Acetic acid}: \quad Y_2O_3 + 2CH_3COOH + H_2O \rightarrow (Y(CH_3COO)(OH)_2)_2 \tag{1}$$

$$\text{Hydrochloric acid}: \quad Y_2O_3 + 3HCl \rightarrow YCl_3 + Y(OH)_3 \tag{2}$$

$$\text{Nitric acid}: \quad Y_2O_3 + 6HNO_3 \rightarrow 2Y(NO_3)_3 + 3H_2O \tag{3}$$

By mixing the catalyst to the yttrium ions obtained from the previous reactions, yttrium hydroxycarbonate is formed (Equation (4)). After microwave irradiation, it is observed the formation of yttrium oxycarbonate (Equation (5)). The powder formed is thus subjected to the final calcination step, in which the yttrium oxide nanostructures are obtained (Equation (6)) [43,45]:

$$2Y^{3+} + 7H_2O + 3CH_4N_2O \rightarrow 2YOHCO_3 + 4NH_4^+ + CO_2 \tag{4}$$

$$2YOHCO_3 \overset{Synthesis}{\rightarrow} Y_2O_2CO_3 + H_2O + CO_2 \tag{5}$$

$$Y_2O_2CO_3 \overset{annealing}{\rightarrow} Y_2O_3 + CO_2 \tag{6}$$

3.1. Structural Characterization of Yb/Eu Doped Y_2O_3 Nanostructures

Figure 2 shows the XRD diffractograms of the nanostructures obtained by microwave synthesis, when using different acids for the dissociations of the primary oxides, and before any heat treatment. It is possible to observe that the as-synthesized nanostructures are amorphous when using hydrochloric and nitric acid. Regarding the use of acetic acid, the synthesized nanostructures present some degree of crystallinity.

To find out the temperature at which there was an amorphous to crystalline phase transformation, *in situ* XRD analyses were carried out in the temperature range from 300 °C to 1000 °C. From the annealing process, in 100 °C steps, it is expected to complete the formation of crystalline Yb/Eu doped Y_2O_3 nanostructures from the intermediate products and sesquioxide phases of $Y_{2-x}O_{3-x}$ [23], as the temperature is raised. On Figure 3, it is possible to observe the XRD diffractograms evolution with the increase of temperature and the respective contour plot, for the Yb/Eu doped Y_2O_3 nanostructures produced using all three acids. It can be observed that for all materials, the complete phase shift to crystalline Y_2O_3 occurs at 700 °C, without expressive changes up to 1000 °C.

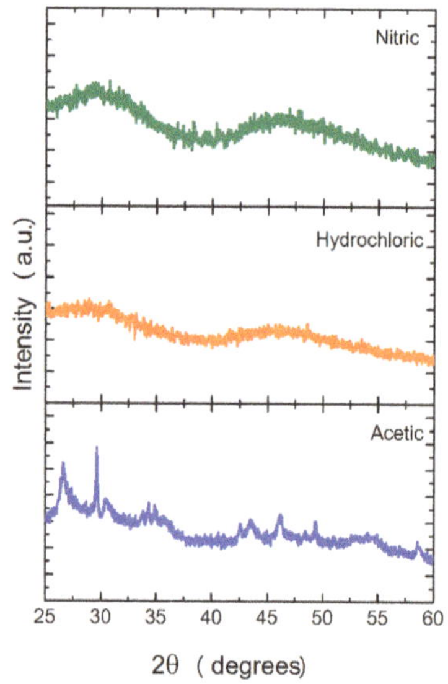

Figure 2. X-Ray diffraction (XRD) diffractograms of nanostructures after microwave synthesis and before any calcination, for the three sets of reactions defined above.

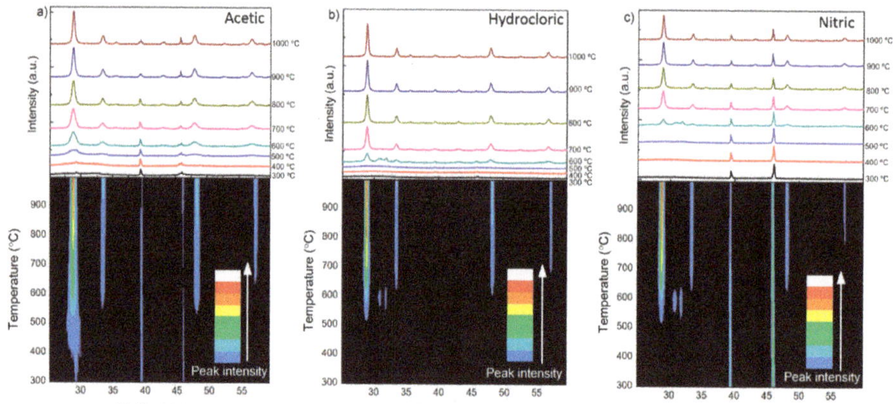

Figure 3. *In situ* XRD diffractograms as a function of temperature (on top) and the respective contour plot (on bottom) of Yb/Eu doped Y_2O_3 nanostructures after microwave synthesis, when using (**a**) acetic, (**b**) hydrochloric and (**c**) nitric acids.

Figure 4 shows the XRD diffractograms of Yb/Eu doped Y_2O_3 nanostructures, prepared with different types of acids, and annealed at 700 °C, for 6 h. For comparison, the XRD simulated Y_2O_3 powder pattern is also shown. It is possible to observe that regardless the used acid, after calcination, it was obtained crystalline Y_2O_3 nanostructures, having a cubic type structure, with the main reflections being (222), (400), (440) and (622), which is in accordance to the literature [23,42]. No peaks shift, or other impurity phases were detected, indicating that a high purity Yb/Eu doped Y_2O_3 nanostructures were obtained by annealing at 700 °C for 6 h.

The crystallite size, D, was estimated using the Scherrer's equation and the most intense peak, corresponding to (222) plane [37]:

$$D = \frac{0.9\,\lambda}{B\cos\theta} \quad (7)$$

where λ is the X-ray wavelength (corresponding to Cu K_α radiation—1.540598 Å), B is the full width at half-maximum (FWHM) in radian and θ is the Bragg diffraction angle degrees. It was obtained a crystallite size of 16.12 nm, 21.88 and 19.86 nm for synthesis with acetic, hydrochloric and nitric acids, respectively.

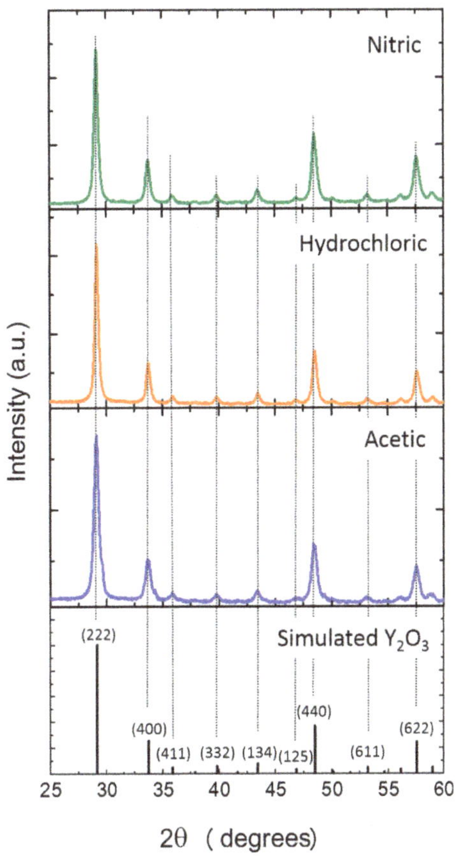

Figure 4. XRD diffractograms of Yb/Eu doped Y_2O_3 nanostructures after calcination at 700 °C. The simulated Y_2O_3 powder pattern is also shown for comparison.

In order to understand the different transformations that occur during Yb/Eu doped Y_2O_3 calcination, differential scanning calorimetric measurements were carried out for all materials. The results are shown in Figure 5. The microwave synthesized materials, which consist in yttrium oxycarbonates were analyzed and it is possible to observe a small mass loss before 110–200 °C, that corresponds to a possible vaporization of the sorbet water. The second weight loss step, below 400–510 °C can be attributed to a dehydration of the yttrium oxycarbotates. The complete conversion into Y_2O_3 occurs at temperatures between 400 and 800 °C corresponding to a mass loss step of 20% for hydrochloric and nitric acid and only 12% for acetic acid. This mass loss is accompanied by an endothermic peak centered at 611–650 °C. Above this temperature, no other weight loss steps or peaks are observed, indicating that the complete conversion of the yttrium oxycarbonate into Y_2O_3 has occurred. These results are in agreement to what is described in literature [45,48] and to what was observed on *in situ* XRD results (Figure 3). These results justify the selected calcination temperature used in the present study, since above 700 °C, no other phase transformation could be detected.

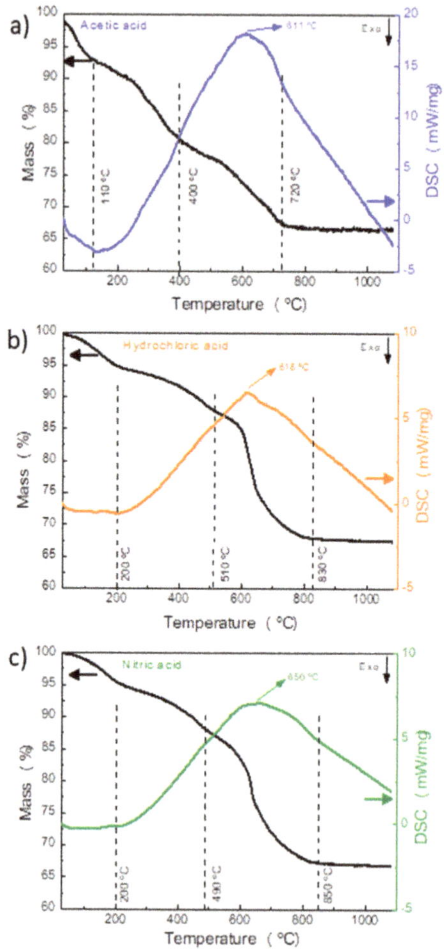

Figure 5. Thermogravimetric analysis and differential scanning calorimetry (TGA/DSC) curves of the as-synthesized nanostructures before calcination produced with (**a**) acetic acid, (**b**) hydrochloric acid and (**c**) nitric acid.

SEM and STEM analyses were carried out for all the materials produced. Figure 6 shows SEM images of the materials after microwave synthesis and before and after calcination. As can be seen, the shape of the nanostructures is maintained after calcination. Nevertheless, after heat exposure, the shape of such structures is better defined, especially for the nitric-based structures. When comparing the three acids used, it is evident that the acetic acid resulted in thin nanosheets, while both hydrochloric and nitric resulted in perfect sphere-like structures (Figures 6 and 7). It is also clear that after calcination, it is observed a reduced diameter that can be explained by sintering, where small primary single crystals diffuse across the boundaries and coalescence to form a larger one. The total volume decreased because a densely packed with the elimination of pores was formed.

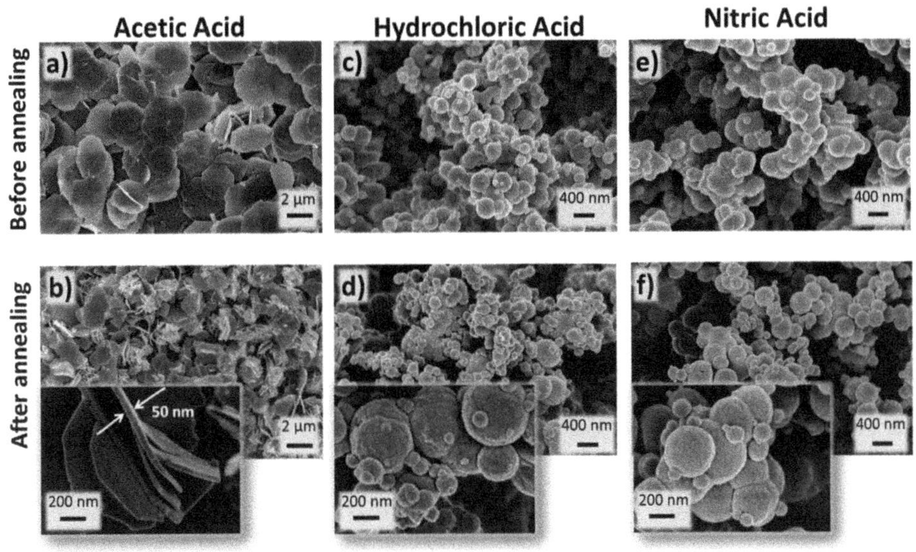

Figure 6. Scanning electron microscopy (SEM) images of Yb/Eu doped Y_2O_3 nanostructures produced by hydrothermal method assisted by microwave radiation, and using (**a**,**b**) acetic acid, (**c**,**d**) hydrochloric acid and (**e**,**f**) nitric acid, before and after annealing at 700 °C, respectively. The insets magnify the structures produced after calcination.

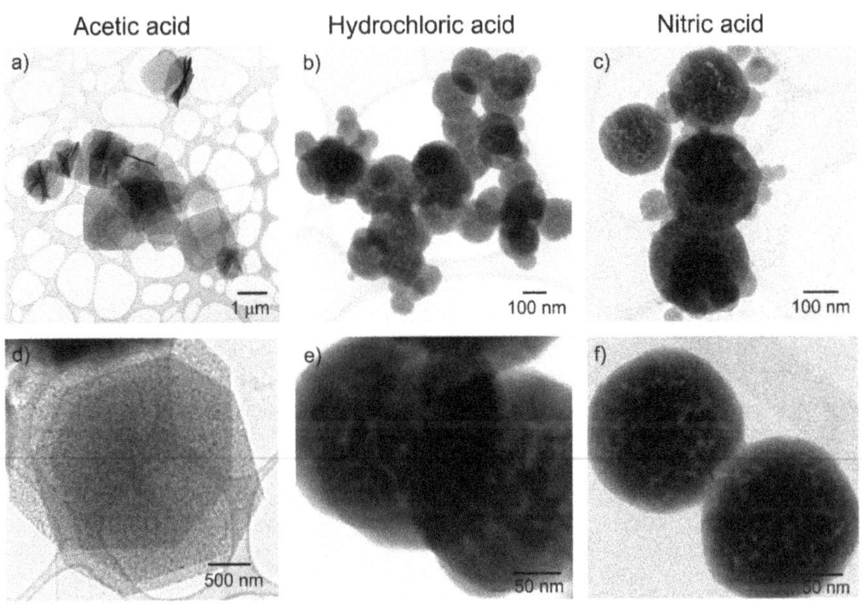

Figure 7. Scanning transmission electron microscopy (STEM) images of Y_2O_3:Yb:Eu nanostructures produced by hydrothermal method assisted by microwave irradiation, and using (**a**,**d**) acetic acid, (**b**,**e**) hydrochloric acid, and (**c**,**f**) nitric acid after calcination at 700 °C.

The nanosheets had an average size of ~2.5 µm and thickness of ~50 nm, while the size of the spheres varied with the acid used. For the hydrochloric acid, the structures had 192 ± 25 nm, and for the nitric, it has been measured an average size of 216 ± 37 nm. Nevertheless, both hydrochloric and nitric-based materials revealed heterogeneities in sphere sizes (this can be observed by the presence of smaller spheres). From Figure 7e,f, it can be observed that the spheres are composed by smaller primary grains, joined to form a compact/granular spherical structure [49]. These smaller grains may justify the crystallite sizes measured by XRD.

EDS measurements were also carried out on both nanostructures formed (see Figure 8). Both nanosheets and spheres showed the presence of Y, Yb, Eu, and O well distributed along the structures. This indicates that the doped Yb^{3+} and Eu^{3+} were successfully incorporated into the Y_2O_3 host matrix. Moreover, these results are in agreement to XRD, where no second phases were detected.

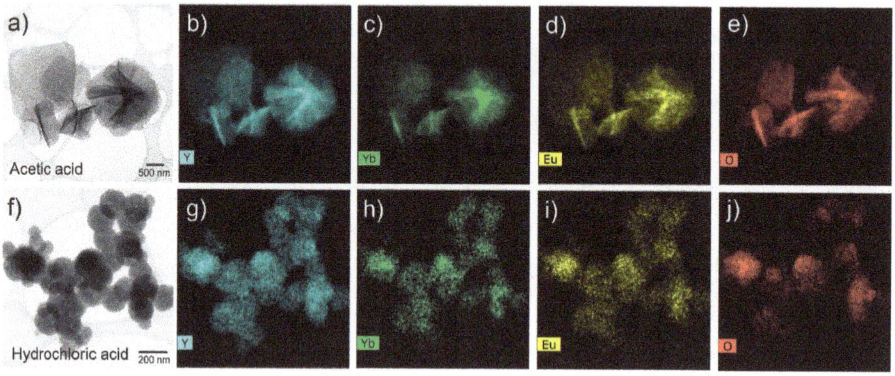

Figure 8. STEM images (**a**,**f**) and energy dispersive X-Ray spectroscopy (EDS) analyses of the Yb/Eu doped Y_2O_3 nanostructures after microwave synthesis and calcination at 700 °C. The corresponding EDS maps for Y (**b**,**g**), Yb (**c**,**h**), Eu (**d**,**i**), and O (**e**,**j**) are presented. The two types of structures have been analyzed.

The possible reasons for these morphological differences in the nanostructures studied are expected to be due to a group of distinct factors. Indeed, the shape of Y_2O_3 nanostructures can be greatly influenced by factors such as intrinsic crystal/cluster structure configuration, precursors, catalysts, surfactants and also by synthesis temperature and time [50]. However, it is believed that the catalyst is responsible for the final nanostructures' morphology. The decomposition of the catalyst into ammonia and carbonates ions (OH^- and $CO_3{}^{2-}$ ions) originates a homogeneous nucleation of Y_2O_3 nanostructures occurring a spontaneous aggregation process in order to minimize the surface energy, forming sphere-shaped nanostructures [50,51]. These spherical nanostructures are an intermediate step in the morphological evolution into nanosheets structures. By increasing the amount of catalyst or by increasing the synthesis time, the dissolution–recrystallization equilibrium will cause the dissolution of smaller particles and the anisotropically grow of larger particles (into sheets) during the Ostwald ripening period [50]. Liu et al. [52] synthesized Y_2O_3:Eu^{3+} nanospheres and by increasing synthesis time while keeping constant the catalyst concentration, they obtained nanoparticles with irregular shape. Khachatourian et al. [51] also studied the effect of catalyst on the morphology of Y_2O_3 nanoparticles. They observed that by increasing the catalyst into a critical point, the diameter nanosphere would decrease (a higher degree of supersaturation would lead into a burst formation of nuclei due to the presence of a higher quantity of OH^- and $CO_3{}^{2-}$ ions, originating a decrease on particle size). Above that point it would result in larger particles [51]. A higher concentration of catalyst will act both as hydrogen-bond donors and acceptors (through C=O groups), resulting in self-association into two-dimensional layered frameworks (sheets nanostructures) [50,53].

By using carboxylic acid (such as acetic acid), it is possible to synthesize a series of nanocrystals connected orthogonally by hydrogen-bonded NH chains and carboxylic acid dimers, and by this way design two-dimensional layered networks [53]. In the case of the other acids tested, hydrochloric and nitric acids resulted in sphere-like structures—but with differences in size—which can be expected since the dissociation reaction of primary oxides should be different regarding the acid used.

3.2. Optical Characterizations

The optical band gaps of Yb/Eu doped Y_2O_3 nanostructures were evaluated from diffuse reflectance data and through the Tauc relation using Equation (8):

$$(\alpha h v)^m = A(hv - Eg) \tag{8}$$

where α is the absorption coefficient, hv is the energy of photons impinging on the material, A is a constant of proportionality function of the matrix density of states and E_g is the optical band gap [54]. For such materials, it has been reported that they exhibit a direct allowed transition [55]. Few studies reported the optical band gaps for analogous materials, nevertheless Halappa et al. [55] described Eu^{3+} activated Y_2O_3 red nanophosphors with a band gap of 5.51 eV. In another study, Shivaramu et al. [56] demonstrated a band gap value of 5.7 eV also to Y_2O_3:Eu^{3+} nanophosphors. In the present study, the band gap values calculated were 4.56 eV, 4.36 eV and 4.53 eV for the materials produced with acetic, hydrochloric and nitric acids, respectively (Figure 9). The values are expressively low, when compared to the literature; however, it is known that adding dopants into the Y_2O_3 lattice causes energy levels in the energy gap between the conduction and valence bands (energy levels of donors or acceptors), which will lead to the decrease of the optical band gap [57]. In fact, Cabello-Guzmán et. al. [57] reported the progressive decrease of optical band gap by having two (Y_2O_3:Er –5.02 eV) or three dopants (Y_2O_3:Er-Yb –4.78 eV) in the host matrix. Moreover, it is expected that material defects and oxygen vacancies can also form discrete energy levels, influencing so the nanostructure's band gaps. The band gap differences observed between the materials can also be justified by the size and structure disparities observed.

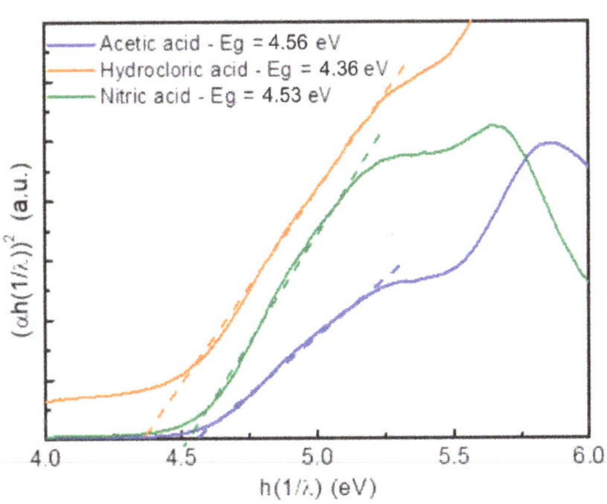

Figure 9. Tauc plot for the Yb/Eu doped Y_2O_3 nanostructures produced with acetic, hydrochloric and nitric acids, respectively.

No direct relation between the nanostructure sizes measured by SEM/STEM and the band gaps values can be stated, however, considering the XRD results (Figure 4), a trend can be inferred, since the lower crystalline size material achieved the highest band gap value (acetate-based material), while the hydrochloric material revealed the lowest band gap value and displayed the highest crystallite size. It is normally accepted that the band gap is strongly dependent on crystallite size [58–60], and that the decrease in the band gap can be related to the increase in the grain size [58]. Moreover, other factors such as the degree of compactness and densification can also contribute to the final band gap value [61].

Figure 10 shows the emission spectra of the Yb^{3+}:Eu^{3+} doped Y_2O_3 nanostructures prepared with different acids and a schematic of the relevant luminescence transitions between the $^5D_0$ and the ^7F_J states [44,62]. The produced materials exhibited a strong red emission under excitation with a 976 nm laser. Nevertheless, it is known that the Eu^{3+} ions cannot be excited with an excitation source around 980 nm due to a considerable mismatch of energy levels in Eu^{3+} ions. However, when co-doped with Yb^{3+}, the upconversion emission occurs, indicating that the Eu^{3+} ions are excited with the Yb^{3+} ions [2,63].

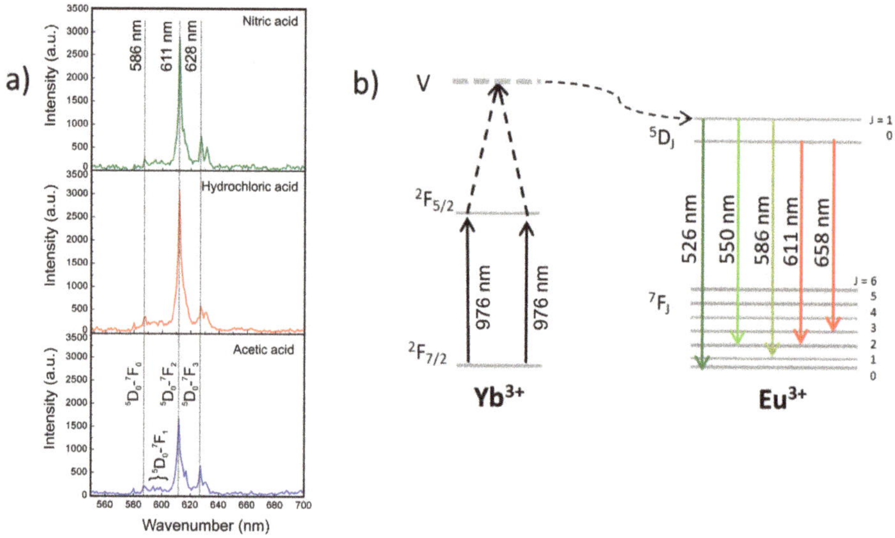

Figure 10. (**a**) Emission spectra of the Yb/Eu doped Y_2O_3 nanostructures obtained at 976 nm, and (**b**) Energy level diagram for Yb^{3+} and Eu^{3+} together with the relevant luminescence transitions between the $^5D_0$ and the ^7F_J states.

When exposed to a 976 nm source, pairs of Yb^{3+} ions are excited from its $^2F_{7/2}$ ground state to the $^2F_{5/2}$. Then, their energy is transferred cooperatively in a way that one of them (acceptor) after gaining energy from the donor, occupies the virtual state (V) (i.e., $^2F_{5/2} + {}^2F_{5/2} \rightarrow 2^2F_{5/2}$). At this point, the excited Yb^{3+} ions from the virtual state transfer their excitation energy into the ground state of Eu^{3+} ions. Therefore, the Eu^{3+} ion is assumed to transit to the $^5D_2$ multiplet term or to the $^5D_1$ multiplet, followed by a nonradiative decay of the $^5D_0$, and thus by emission [2,63].

The emission spectra observed in Figure 10 revealed radiative emissions at ~586 nm, 611 nm, and 628 nm, for all the materials produced. The emission bands can be assigned to the $^5D_0 \rightarrow {}^7F_J$ (J = 0, 1, 2) transitions of Eu^{3+} [63]. The strongest peak at 611 nm is originated from $^5D_0$ to $^7F_2$ transition. When comparing between the material's emission intensities, it can be observed that the emission bands are sharpen for the hydrochloric and nitric acid-based materials (the maximum values for the $^5D_0$–$^7F_2$ transitions were 3050 and 2840, for hydrochloric and nitric acid-based materials, respectively). The lower emission observed for the acetic-based material (maximum value of 1650 for the $^5D_0$–$^7F_2$ transition) can be associated to a size effect, since the luminescent intensity is known to be inversely proportional to the particle size, and thus the larger sized nanosheets are expected to have lower luminescence. Nevertheless, several aspects may influence the relative emission intensities of all materials, including absorption cross-section, competition between radiative cooperative emission, energy transfer to Eu^{3+} and energy migration to traps [63], besides a mesoscopic effect, associated to the shape and configuration of the nanostructures produced.

4. Conclusions

The fast microwave syntheses (15 min) of Yb/Eu doped Y_2O_3 nanostructures was reported. Microwave syntheses revealed itself to be an efficient way for successfully doping the Y_2O_3 host matrix after dissociating the primary oxides with different acids. In fact, this study revealed the key role of the acids used regarding the final structure and optical behaviour of the nanostructures produced. SEM and STEM revealed that the acetic acid resulted in nanosheets, while both hydrochloric and nitric formed nanospheres with heterogeneities in size. The study indicated the calcination temperature (700 °C) at which the conversion into Y_2O_3 is complete for such nanostructures. This temperature range is expected to cause an impact on production and energy costs of such nanostructures—in fact the calcination temperature is substantially lower than what is usually reported in literature. The optical properties of the produced materials were discussed, and the photoluminescence experiments showed a red Eu^{3+} emission when exited to a 976 nm source. Moreover, the sphere-like structures revealed enhanced luminescence, when compared to the nanosheets. The lower emission of the nanosheets was associated to a size effect, since acetic acid originated larger structures. The present study opens up to the possibility of these materials to be used as infrared to visible upconverters, and their potential to be integrated in several opto-electronic devices. For further studies, it is imperative to control the particle size and the size distribution of the developed nanostructures, which can be achieved by altering the catalyst or its concentration, but also with further investigation and adjustments on the synthesis parameters.

Author Contributions: D.N. and A.P. were responsible for writing the original manuscript with equal contribution; M.M. and T.F. performed the experiments; optical characterization was performed by A.P.; the DSC-TG and XRD characterization was performed by A.P.; D.N. performed the SEM; P.A.C. performed the STEM measurements; A.A., F.S., P.G. and S.G. participated in the review and editing and are responsible for funding, and finally the paper was under the supervision of R.M. and E.F.

Funding: This work was supported by FEDER funds, through the COMPETE 2020 Program, and national funds, through the Fundação para Ciência e Tecnologia (FCT), under the projects POCI-01-0145-FEDER-007688 (Reference UID/CTM/50025). The authors also acknowledge funding from the European Commission through the projects 1D-NEON (H2020-NMP-2015, grant 685758-21D) and BET-EU (H2020-TWINN-2015, grant 692373. The work was also partially funded by the Nanomark collaborative project between INCM (Imprensa Nacional—Casa da Moeda) and CENIMAT/i3N. D. Nunes and A. Pimentel acknowledge funding from FCT through the grants SFRH/BPD/84215/2012 and SFRH/BPD/76992/2011, respectively. P.A. Carvalho acknowledges support from the Research Council of Norway through grants 275752 and 197405/F50.

Conflicts of Interest: The authors declare no conflicts of interest.

References

1. De Wild, J.; Meijerink, A.; Rath, J.K.; Van Sark, W.G.J.H.M.; Schropp, R.E.I. Upconverter solar cells: Materials and applications. *Energy Environ. Sci.* **2011**, *4*, 4835–4848. [CrossRef]
2. Pandey, A. Sensitization Effect of Yb^{3+} in Upconversion Luminescence of Eu^{3+}Codoped Y_2O_3 Phosphor. *J. Phys. Chem. Biophys.* **2013**, *3*, 1–3. [CrossRef]
3. Nazarov, M.; Noh, D.Y. Rare earth double activated phosphors for different applications. *J. Rare Earths* **2010**, *28*, 1–11. [CrossRef]
4. Atwood, D.A. *The Rare Earth Elements: Fundamentals and Application*; John Wiley & Sons: New York, NY, USA, 2006. [CrossRef]
5. Bünzli, J.C.G.; Eliseeva, S.V. Lanthanide NIR luminescence for telecommunications, bioanalyses and solar energy conversion. *J. Rare Earths* **2010**, *28*, 824–842. [CrossRef]
6. Bünzli, J.-C.G.; Eliseeva, S.V. Basics of Lanthanide Photophysics. *Lanthan. Lumin.* **2010**, *7*, 1–45. [CrossRef]
7. Bedekar, V.; Dutta, D.P.; Mohapatra, M.; Godbole, S.V.; Ghildiyal, R.; Tyagi, A.K. Rare-earth doped gadolinia based phosphors for potential multicolor and white light emitting deep UV LEDs. *Nanotechnology* **2009**, *20*, 125707. [CrossRef] [PubMed]
8. Guo, C.; Huang, D.; Su, Q. Methods to improve the fluorescence intensity of CaS:Eu^{2+} red-emitting phosphor for white LED. *Mater. Sci. Eng. B* **2006**, *130*, 189–193. [CrossRef]

9. Ballato, J.; Lewis, J.S.; Holloway, P. Display Applications of Rare-Earth-Doped Materials. *MRS Bull.* **1999**, *24*, 51–56. [CrossRef]
10. Leskelä, M.; Niinistö, L. Applications of rare earths in full-colour EL displays. *Mater. Chem. Phys.* **1992**, *31*, 7–11. [CrossRef]
11. Kenyon, A. Recent developments in rare-earth doped materials for optoelectronics. *Prog. Quantum Electron.* **2002**, *26*, 225–284. [CrossRef]
12. Optical Sensor Utilizing Rare-Earth-Doped Integrated-Optic Lasers. 1995. Available online: https://patents.google.com/patent/US5677769A/en (accessed on 28 November 2018).
13. Steckl, A.J.; Zavada, J.M. Photonic Applications of Rare-Earth-Doped Materials. *MRS Bull.* **1999**, *24*, 16–20. [CrossRef]
14. Rare Earth Doped and Large Effective Area Optical Fibers for Fiber Lasers and Amplifiers. 2010. Available online: https://patents.google.com/patent/US8213758B2/en (accessed on 28 November 2018).
15. Seddon, A.B.; Tang, Z.; Furniss, D.; Sujecki, S.; Benson, T.M. Progress in rare-earth-doped mid-infrared fiber lasers. *Opt. Express* **2010**, *18*, 26704. [CrossRef] [PubMed]
16. Zhou, X.; Huang, Y.; Liu, C.; Liao, J.; Lu, T.; Xing, W. Available hydrogen from formic acid decomposed by rare earth elements promoted Pd-Au/C catalysts at low temperature. *ChemSusChem* **2010**, *3*, 1379–1382. [CrossRef] [PubMed]
17. Yu, L.; Lu, Y.; Man, N.; Yu, S.-H.; Wen, L.-P. Rare Earth Oxide Nanocrystals Induce Autophagy in HeLa Cells. *Small* **2009**, *5*, 2784–2787. [CrossRef] [PubMed]
18. Xu, C.; Qu, X. Cerium oxide nanoparticle: A remarkably versatile rare earth nanomaterial for biological applications. *NPG Asia Mater.* **2014**, *6*, e90. [CrossRef]
19. Shalav, A.; Richards, B.S.; Green, M.A. Luminescent layers for enhanced silicon solar cell performance: Up-conversion. *Sol. Energy Mater. Sol. Cells* **2007**, *91*, 829–842. [CrossRef]
20. Vicente, A.T.; Araújo, A.; Mendes, M.J.; Nunes, D.; Oliveira, M.J.; Sanchez-Sobrado, O.; Ferreira, M.P.; Águas, H.; Fortunato, E.; Martins, R. Multifunctional cellulose-paper for light harvesting and smart sensing applications. *J. Mater. Chem. C* **2018**, *6*, 3143–3181. [CrossRef]
21. Hu, S.; Zhang, H.; Yang, J.; Qiao, Z. Hydrothermal Synthesis and Green Luminescent Properties of Lu_2O_3:Yb^{3+}/Ho^{3+} Nanocubes. *J. Nanosci. Nanotechnol.* **2014**, *14*, 3853–3860. [CrossRef]
22. Jones, S.L.; Kumar, D.; Singh, R.K.; Holloway, P.H. Luminescence of pulsed laser deposited Eu doped yttrium oxide films. *Appl. Phys. Lett.* **1998**, *71*, 404. [CrossRef]
23. Hao, J.; Studenikin, S.A.; Cocivera, M. Blue, green and red cathodoluminescence of Y_2O_3 phosphor films prepared by spray pyrolysis. *J. Lumin.* **2001**, *93*, 313–319. [CrossRef]
24. Chen, K.; Peng, J.; Srinivasakannan, C.; Yin, S.; Guo, S.; Zhang, L. Effect of temperature on the preparation of yttrium oxide in microwave field. *J. Alloys Compd.* **2018**, *742*, 13–19. [CrossRef]
25. Mangalaraja, R.V.; Mouzon, J.; Hedström, P.; Camurri, C.P.; Ananthakumar, S.; Odén, M. Microwave assisted combustion synthesis of nanocrystalline yttria and its powder characteristics. *Powder Technol.* **2009**, *191*, 309–314. [CrossRef]
26. Agrawal, S.; Dubey, V. Down conversion luminescence behavior of Er and Yb doped Y_2O_3 phosphor. *J. Radiat. Res. Appl. Sci.* **2014**, *7*, 601–606. [CrossRef]
27. Mishra, K.C.; Berkowitz, J.K.; Johnson, K.H.; Schmidt, P.C. Electronic structure and optical properties of europium-activated yttrium oxide phosphor. *Phys. Rev. B* **1992**, *45*, 10902–10906. [CrossRef]
28. Chávez, D.H.; Contreras, O.E.; Hirata, G.A. Synthesis and Upconversion Luminescence of Nanoparticles Y_2O_3 and Gd_2O_3 Co-doped with Yb^{3+} and Er^{3+}. *Nanomater. Nanotechnol.* **2016**, *6*, 7. [CrossRef]
29. Yang, J.; Gu, J.; Yang, R.; Shang, Q.; Yang, J. Hydrothermal Synthesis Y_2O_3:Yb^{3+}/Er^{3+} Nanospheres with Upconversion Luminescence from Green to Red. *Nanosci. Nanometrol.* **2016**, *2*, 41–45. [CrossRef]
30. Dhanaraj, J.; Jagannathan, R.; Kutty, T.R.N.; Lu, C.-H. Photoluminescence Characteristics of Y_2O_3:Eu^{3+} Nanophosphors Prepared Using Sol−Gel Thermolysis. *J. Phys. Chem. B* **2001**, *105*, 11098–11105. [CrossRef]
31. Zhai, Y.; Yao, Z.; Ding, S.; Qiu, M.; Zhai, J. Synthesis and characterization of Y_2O_3:Eu nanopowder via EDTA complexing sol–gel process. *Mater. Lett.* **2003**, *57*, 2901–2906. [CrossRef]
32. Lakshminarasappa, B.N.; Jayaramaiah, J.R.; Nagabhushana, B.M. Thermoluminescence of combustion synthesized yttrium oxide. *Powder Technol.* **2012**, *217*, 7–10. [CrossRef]

33. Capobianco, J.A.; Vetrone, F.; D'Alesio, T.; Tessari, G.; Speghini, A.; Bettinelli, M. Optical spectroscopy of nanocrystalline cubic Y_2O_3:Er^{3+} obtained by combustion synthesis. *Phys. Chem. Chem. Phys.* **2000**, *2*, 3203–3207. [CrossRef]
34. Chen, J.; Huang, B.; Huang, C.; Sun, X. Preparation of nanoscaled yttrium oxide by citrate precipitation method. *J. Rare Earths* **2017**, *35*, 79–84. [CrossRef]
35. Jeong, J.Y.; Park, S.W.; Moon, D.K.; Kim, W.J. Synthesis of Y_2O_3 nano-powders by precipitation method using various precipitants and preparation of high stability dispersion for backlight unit (BLU). *J. Ind. Eng. Chem.* **2010**, *16*, 243–250. [CrossRef]
36. Tian, Y.; Lu, F.; Xing, M.; Ran, J.; Fu, Y.; Peng, Y.; Luo, X. Upconversion luminescence properties of Y_2O_2S:Er^{3+}@Y_2O_2S:Yb^{3+}, Tm^{3+} core-shell nanoparticles prepared via homogeneous co-precipitation. *Opt. Mater. (Amst.)* **2017**, *64*, 58–63. [CrossRef]
37. Khachatourian, A.M.; Golestani-Fard, F.; Sarpoolaky, H.; Vogt, C.; Vasileva, E.; Mensi, M.; Popov, S.; Toprak, M.S. Microwave synthesis of Y_2O_3:Eu^{3+} nanophosphors: A study on the influence of dopant concentration and calcination temperature on structural and photoluminescence properties. *J. Lumin.* **2016**, *169*, 1–8. [CrossRef]
38. Malek Khachatourian, A.; Golestani-Fard, F.; Sarpoolaky, H.; Vogt, C.; Toprak, M.S. Microwave assisted synthesis of monodispersed Y_2O_3 and Y_2O_3:Eu^{3+} particles. *Ceram. Int.* **2015**, *41*, 2006–2014. [CrossRef]
39. Pimentel, A.; Samouco, A.; Nunes, D.; Araújo, A.; Martins, R.; Fortunato, E. Ultra-fast microwave synthesis of ZnO nanorods on cellulose substrates for UV sensor applications. *Materials (Basel)* **2017**, *10*, 1308. [CrossRef]
40. Pimentel, A.; Nunes, D.; Pereira, S.; Martins, R.; Fortunato, E. Photocatalytic Activity of TiO_2 nanostructured arrays prepared by microwave-assisted solvothermal method. In *Semiconductor Photocatalysis—Materials, Mechanisms and Applications*; Cao, W., Ed.; InTech: Rijeka, Croatia, 2016. [CrossRef]
41. Nunes, D.; Pimentel, A.; Barquinha, P.; Carvalho, P.A.; Fortunato, E.; Martins, R. Cu_2O polyhedral nanowires produced by microwave irradiation. *J. Mater. Chem. C* **2014**, *2*, 6097. [CrossRef]
42. Anh, T.; Benalloul, P.; Barthou, C.; Giang, L.T.; Vu, N.; Minh, L. Luminescence, energy transfer, and upconversion mechanisms of Y_2O_3 nanomaterials doped with Eu^{3+}, Tb^{3+}, Tm^{3+}, Er^{3+}, and Yb^{3+} ions. *J. Nanomater.* **2007**, *2007*, 1–10. [CrossRef]
43. Sordelet, D.; Akinc, M. Preparation of spherical, monosized Y_2O_3 precursor particles. *J. Colloid Interface Sci.* **1988**, *122*, 47–59. [CrossRef]
44. Jia, G.; Yang, M.; Song, Y.; You, H.; Zhang, H. General and Facile Method To Prepare Uniform Y_2O_3:Eu Hollow Microspheres. *Cryst. Growth Des.* **2009**, *9*, 301–307. [CrossRef]
45. Sung, J.M.; Lin, S.E.; Wei, W.C.J. Synthesis and reaction kinetics for monodispersive Y_2O_3:Tb^{3+} spherical phosphor particles. *J. Eur. Ceram. Soc.* **2007**, *27*, 2605–2611. [CrossRef]
46. Schneider, C.A.; Rasband, W.S.; Eliceiri, K.W. NIH Image to ImageJ: 25 years of image analysis. *Nat. Methods* **2012**, *9*, 671–675. [CrossRef]
47. Byrappa, K.; Yoshimura, M. *Handbook of Hydrothermal Technology—A Technology for Crystal Growth and Materials Processing*; Noyes Publication: Norwich, NY, USA, 2001.
48. Balabanov, S.S.; Gavrishchuk, E.M.; Permin, D.A. Synthesis and properties of yttrium hydroxyacetate sols. *Inorg. Mater.* **2012**, *48*, 500–503. [CrossRef]
49. Lojpur, V.; Nikolic, M.; Mancic, L.; Milosevic, O.; Dramicanin, M.D. Y_2O_3:Yb,Tm and Y_2O_3:Yb,Ho powders for low-temperature thermometry based on up-conversion fluorescence. *Ceram. Int.* **2013**, *39*, 1129–1134. [CrossRef]
50. Wang, H.; Uehara, M.; Nakamura, H.; Miyazaki, M.; Maeda, H. Synthesis of well-dispersed Y_2O_3:Eu nanocrystals and self-assembled nanodisks using a simple non-hydrolytic route. *Adv. Mater.* **2005**, *17*, 2506–2509. [CrossRef]
51. Sarpoolaky, H.; Fard, F.G.; Khachatourian, A.M. Ceramics International Microwave Assisted Synthesis of Monodispersed Y_2O_3 and Y_2O_3:Eu3þ Particles. Available online: http://www.academia.edu/36133797/CERAMICS_INTERNATIONAL_Microwave_assisted_synthesis_of_monodispersed_Y_2_O_3_and_Y_2_O_3_Eu_3_\T1\th_particles (accessed on 22 December 2018).
52. Liu, W.; Wang, Y.; Zhang, M.; Zheng, Y. Synthesis of Y_2O_3:Eu^{3+} coated Y_2O_3 phosphors by urea-assisted homogeneous precipitation and its photoluminescence properties. *Mater. Lett.* **2013**, *96*, 42–44. [CrossRef]
53. Custelcean, R. Crystal engineering with urea and thiourea hydrogen-bonding groups. *Chem. Commun.* **2008**, *3*, 295–307. [CrossRef]

54. Tauc, J. Optical properties and electronic structure of amorphous Ge and Si. *Mater. Res. Bull.* **1968**, *3*, 37–46. [CrossRef]
55. Halappa, P.; Raj, S.T.; Sairani, R.; Joshi, S.; Madhusudhana, R.; Shivakumara, C. Combustion synthesis and characterisation of Eu^{3+}-activated Y_2O_3 red nanophosphors for display device applications. *Int. J. Nanotechnol.* **2017**, *14*, 833–844. [CrossRef]
56. Shivaramu, N.J.; Lakshminarasappa, B.N.; Nagabhushana, K.R.; Singh, F.; Swart, H.C. Synthesis, thermoluminescence and defect centres in Eu^{3+} doped Y_2O_3 nanophosphor for gamma dosimetry applications. *Mater. Res. Express* **2017**, *4*, 115033. [CrossRef]
57. Cabello-Guzmán, G.; González, D.; Caro-Díaz, C.; Lillo-Arroyo, L.; Valenzuela-Melgarejo, F.; Cárdenas Triviño, G.; Buono-Core, G.E.; Chornik, B.; Huentupil, Y. Preliminary evaluation of the up-conversion emission of Y_2O_3:Er-Yb thin films prepared by a solid state photochemical deposition method. *J. Lumin.* **2018**, *204*, 401–409. [CrossRef]
58. Ramana, C.V.; Smith, R.J.; Hussain, O.M. Grain size effects on the optical characteristics of pulsed-laser deposited vanadium oxide thin films. *Phys. Status Solidi* **2003**, *199*, R4–R6. [CrossRef]
59. Migas, D.B.; Filonov, A.B.; Borisenko, V.E.; Skorodumova, N.V. Orientation effects in morphology and electronic properties of anatase TiO_2 one-dimensional nanostructures. II. Nanotubes. *Phys. Chem. Chem. Phys.* **2014**, *16*, 9490–9498. [CrossRef] [PubMed]
60. Nunes, D.; Pimentel, A.; Pinto, J.V.; Calmeiro, T.R.; Nandy, S.; Barquinha, P.; Pereira, L.; Carvalho, P.A.; Fortunato, E.; Martins, R. Photocatalytic behavior of TiO_2 films synthesized by microwave irradiation. *Catal. Today* **2016**, *278*, 262–270. [CrossRef]
61. Pereira, L.; Barquinha, P.; Fortunato, E.; Martins, R.; Kang, D.; Kim, C.J.; Lim, H.; Song, I.; Park, Y. High k dielectrics for low temperature electronics. *Thin Solid Films* **2008**, *516*, 1544–1548. [CrossRef]
62. Xiao, Y.; Han, G.; Chang, Y. Electrospun ytterbium and europium ions co-doped stannic oxide nanofibers and application in dye-sensitized solar cells. *Mater. Res. Bull.* **2017**, *92*, 90–98. [CrossRef]
63. Wang, H.; Duan, C.K.; Tanner, P.A. Visible upconversion luminescence from Y_2O_3:Eu^{3+},Yb^{3+}. *J. Phys. Chem. C* **2008**, *112*, 16651–16654. [CrossRef]

nanomaterials

MDPI

Article

Recovery of $Li(Ni_{0.33}Mn_{0.33}Co_{0.33})O_2$ from Lithium-Ion Battery Cathodes: Aspects of Degradation

Tim Sieber, Jana Ducke, Anja Rietig, Thomas Langner and Jörg Acker *

Department of Physical Chemistry, Brandenburg Technical University Cottbus-Senftenberg, Universitätsplatz 1, D-01968 Senftenberg, Germany; tim.sieber@b-tu.de (T.S.); jana.ducke@b-tu.de (J.D.); anja.rietig@b-tu.de (A.R.); thomas.langner@b-tu.de (T.L.)

* Correspondence: joerg.acker@b-tu.de; Tel.: +49-3573-85-839

Received: 19 December 2018; Accepted: 7 February 2019; Published: 12 February 2019

Abstract: Nickel–manganese–cobalt oxides, with $LiNi_{0.33}Mn_{0.33}Co_{0.33}O_2$ (NMC) as the most prominent compound, are state-of-the-art cathode materials for lithium-ion batteries in electric vehicles. The growing market for electro mobility has led to a growing global demand for Li, Co, Ni, and Mn, making spent lithium-ion batteries a valuable secondary resource. Going forward, energy- and resource-inefficient pyrometallurgical and hydrometallurgical recycling strategies must be avoided. We presented an approach to recover NMC particles from spent lithium-ion battery cathodes while preserving their chemical and morphological properties, with a minimal use of chemicals. The key task was the separation of the cathode coating layer consisting of NMC, an organic binder, and carbon black, from the Al substrate foil. This can be performed in water under strong agitation to support the slow detachment process. However, the contact of the NMC cathode with water leads to a release of Li^+ ions and a fast increase in the pH. Unwanted side reactions may occur as the Al substrate foil starts to dissolve and $Al(OH)_3$ precipitates on the NMC. These side reactions are avoided using pH-adjusted solutions with sufficiently high buffer capacities to separate the coating layer from the Al substrate, without precipitations and without degradation of the NMC particles.

Keywords: lithium-ion; nickel–manganese–cobalt oxide (NMC); leaching; recycling; recover; degradation; SEM-EDX; Raman spectroscopy

---

1. Introduction

The expanding market for electric vehicles requires lithium-ion batteries (Li-ion batteries), as these are energy storage devices with high power, high capacity, high charging rates, and long life stability. The most common cathode materials for Li-ion batteries in electric vehicles [1] belong to a group of layered mixed transition metal oxide compounds with rhombohedra symmetry (D_{3d}^5 space group), $LiNi_xMn_yCo_{1-x-y}O_2$ (with $(x+y) \in [0,1]$) [2]. Among them, the most popular and widely used material in Li-ion batteries is the phase $LiNi_{0.33}Mn_{0.33}Co_{0.33}O_2$ (NMC).

Global demand for powerful, rechargeable Li-ion batteries, particularly for electric vehicles, has increased the demand for essential elements Co, Ni, Mn, and Li. For years, Co has been the most valuable and critical raw element needed in battery metals [3]. In 2005, 25% of the end products made with cobalt in the EU were used in the manufacturing of battery chemicals, and this value had increased to 42% in 2015 [4]. Therefore, end-of-life Li-ion batteries have become an essential secondary source to cover the Co requirements and for other necessary elements.

There are two basic technologies for the recycling of spent Li-ion batteries: The pyrometallurgical and the hydrometallurgical process [5–14]. The pyrometallurgical route is a smelting process in which spent Li-ion batteries are entirely melted down without further pretreatment. Organic materials are needed as fuel to maintain the melting process. A molten metal phase consisting of Co, Cu, and Ni is obtained from the smelter and leached in acid solution after cooling. After separation and chemical precipitation, the metals are recovered as inorganic salts. Al, the substrate material of the cathode, Li, and Mn are oxidized, and they are combined with other metal oxides as a slag that can be treated to recover Li. Recently, a combined route consisting of mechanical treatment (comminution of the cathodes and mechanical separation of the Al substrate foils and NMC) and hydrometallurgical treatment (dissolution of the NMC and metal recovery therefrom) was described [14]. The pyrometallurgical route is energy-intense due to the high temperatures required. It emits dust and hazardous gaseous compounds and results in a significant loss of materials. The hydrometallurgical route comprises the leaching of dismounted or shredded Li-ion batteries in strong inorganic acids to dissolve any metals and battery materials. The major challenge in this process is the separation of the ions from the concentrated metal ion liquor, mainly by selective precipitation, electrochemical deposition, or other techniques, such as solvent extraction or ion exchange. As in the pyrometallurgical route, the metals are obtained as inorganic salts. Losses typically occur from insufficient leaching and the precipitation/separation efficiencies of the metal salts.

Our motivation was to avoid the pyrometallurgical and hydrometallurgical recycling routes, as well as their specific drawbacks. In this paper, we described the basics of an alternative approach to recover NMC while preserving its chemical, physical, and morphological properties, with a minimal use of chemicals. This approach, which is designated as functional recycling (Figure 1), can be applied to the cathodes of dismantled and separated end-of-life Li-ion batteries, as well as to residues or scrap from the production of cathodes. Based on the known design of a cathode in Li-ion batteries (Figure 1), the first step in functional recycling is the complete removal of the cathode coating (which consists of NMC, a binder, and conductive carbon black) from the Al substrate. This requires arranging for the medium to be as minimally reactive as possible toward the NMC or the Al substrate foil, and to keep the contact time with the medium as short as possible, so that the NMC particles do not experience any degradation. The development of such a procedure was presented in this work. Basically, this treatment is followed by a second step, not considered here, in which the mechanically separated coating is dried and mechanically comminuted to release and separate the NMC particles from the binder/carbon black mixture.

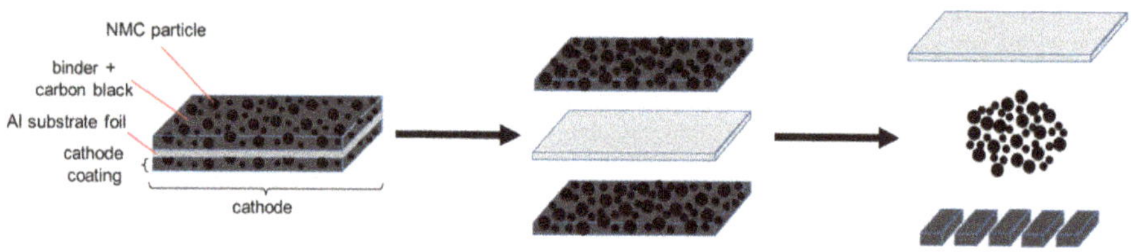

Figure 1. Scheme for the functional recycling of cathodes from Li-ion batteries to recover $LiNi_{0.33}Mn_{0.33}Co_{0.33}O_2$ (NMC) for re-use.

2. Materials and Methods

NMC material and cathodes: The chemical composition of the NMC material used in this study was determined by an inductively coupled plasma-optical emission spectrometer (ICP-OES) to be $Li_{0.945\pm0.007}Ni_{0.336\pm0.001}Mn_{0.331\pm0.001}Co_{0.333\pm0.001}O_2$, which was close to the ideal NMC stoichiometry. The NMC consisted of spherical agglomerates (c.f. Section 3.1), designated as secondary particles, with diameters between 5 µm and 15 µm, which were constituted from primary particles with diameters of 0.5–1 µm (c.f. Section 3.1). The studied cathodes consisted of 20 µm thick Al foils, and they were coated with a layer consisting of NMC, the organic binder polyvinylidene fluoride (PVDF), and carbon black as a conductive additive. Two different cathodes were studied with a mean single-sided coating layer thickness of 50 µm and 22 µm, and they were designated as cathode 1 and cathode 2, respectively.

For cathode 1, the mass fraction of Al amounted to 14.1% of the total weight, and the total Li content was 56.84 mg per 1 g cathode. The mass fraction of Al in cathode 2 amounted to 27.4%, and the total Li content was 45.64 mg per 1 g cathode. A mechanically shredded fraction, as well as a preselected fraction of the cathodes with a mass fraction of 7.0% Al, were used.

Chemicals and reagents: All acids (HCl, HNO_3) and chemicals (citric acid $C_6H_8O_7$, KH_2PO_4, K_2HPO_4, NaOH, Na_2CO_3, and $NaHCO_3$) used in this study were of analytical grade and were purchased from Merck (Darmstadt, Germany). Solutions of $C_6H_8O_7$, KH_2PO_4, K_2HPO_4, and NaOH were prepared by dissolving the chemicals in deionized water (pH between 5.8 and 6.5, 18MΩ cm^{-1}, Milli-Q, Darmstadt, Germany). The $H_2PO_4^-/HPO_4^{2-}$ buffer solutions were prepared by mixing solutions of KH_2PO_4 (c = 66 mmol L^{-1}) and K_2HPO_4 (c = 66 mmol L^{-1}). Li, Ni, Mn, Co, Al, and S single-element standard solutions from Merck were used to prepare multielement calibration standards for ICP-OES measurements. The calibration standards for the measurements by ion chromatography (IC) were prepared from an SO_4^{2-} standard solution from Merck.

Leaching experiments: The cathode foils were gently cut with scissors into square pieces with a size of 5 × 5 mm^2. Pieces with a damaged coating layer or uneven edges were rejected. Weighed samples of the NMC (the cathode pieces or the shredded fraction) were put in contact with the leaching media. If not otherwise stated, the experiments were performed at 25 °C with a 50 mL volume of the leaching medium. All the leaching experiments were stirred at 400 min^{-1} using a magnetic stirrer (H+P Labortechnik AG, Oberschleißheim, Germany). During leaching, suspension samples were collected, and after filtration, the metal concentrations were analyzed by ICP-OES.

Sample digestion: The NMC starting material was dissolved in a mixture of HNO_3, 69% (m/m), and HCl, 37% (m/m), using a high-pressure microwave digestion system (MLS Ethos Start, Leutkirch, Germany, 100 mL polytetrafluoroethylene (PTFE) vessels, maximum temperature 220 °C), and they were subsequently analyzed by ICP-OES.

Chemical analysis: An inductively coupled plasma-optical emission spectrometer (ICP-OES) with a dual-view option (iCap 6500 DUO, Thermo Scientific, Dreieich, Germany) was used to determine the elemental composition of the NMC and the concentrations of the dissolved elements. The sample introduction system was equipped with a parallel path nebulizer made of polyether ether ketone (PEEK) (MiraMist, Burgener Inc., Mississauga, ON, Canada), a cyclonic spray chamber (Glass Expansion, Port Melbourne, Victoria, Australia), and a ceramic injection tube (Glass Expansion, Port Melbourne, Victoria, Australia).

Ion analysis: Concentrations of sulfate ions were determined using ion chromatography (881 compact IC pro, Deutsche METROHM GmbH & Co. KG, Filderstadt, Germany). An METROSEP ASupp5 (Deutsche METROHM GmbH & Co. KG, Filderstadt, Germany; length 250 mm, diameter 4.0 mm) was used as a separation column, with a carbonate eluent consisting of 3.2 mmol L^{-1} Na_2CO_3 and 1.0 mmol L^{-1} $NaHCO_3$. Samples and standard solutions were purified from metal ions using an SPE-H^+-cartridge (MACHEREY-NAGEL GmbH & Co. KG, Düren, Germany).

Raman confocal microscopy: NMC particles were studied with a confocal Raman microscope (DXR SmartRaman, Thermo Fisher Scientific, Dreieich, Germany) in the backscattering configuration. The microscope was equipped with a 532 nm excitation laser and a 900 grooves/mm grating to record the Raman spectra in the wavenumber range of 150–1250 cm^{-1}. The incident laser light, generated with a laser power of 0.5 mW, was focused on the sample surface through a 100× microscope objective. The laser spot had a diameter of 1.6 µm.

SEM-EDX measurements: The materials were studied with SEM-EDX (scanning electron microscopy energy-dispersive X-ray spectroscopy, using a ZEISS EVO M 15 (Carl Zeiss Microscopy GmbH, Jena, Germany) equipped with EDAX TEAM™ EDS (AMETEK, Weiterstadt, Germany).

3. Results

3.1. *Cathode Foil in Contact with Water*

The NMC particles in this study had a typical size between 5 μm and 15 μm (Figure 2a,b) and were processed into coatings with thicknesses typically between 22 μm and 50 μm. Figure 2c shows a SEM micrograph of a cross-section of the studied cathodes. The thickness of the Al substrate, the thickness of the coating, and the NMC content in the coating layer may vary depending on the production process.

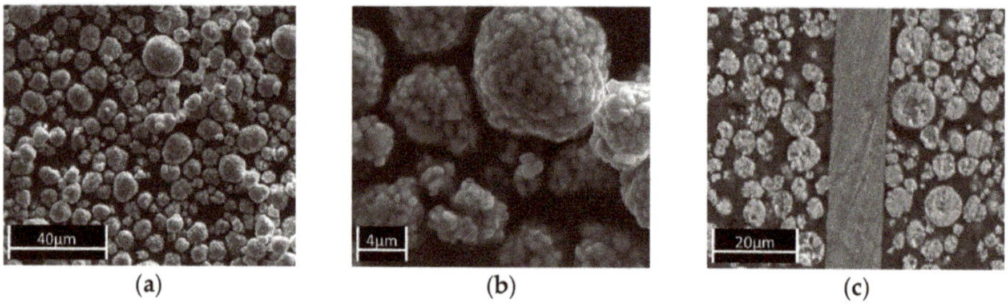

Figure 2. SEM micrographs of the NMC starting material: (**a**) Magnification 1000; (**b**) Magnification 2500; (**c**) Cross section of cathode foil 1.

If a cathode is brought into contact with water, the foil may remain unchanged for days. However, if the cathode foil was mechanically comminuted or cut into square pieces with defined areas between 10 mm^2 and 30 mm^2, and if the aqueous suspension was strongly stirred at speeds between 400 min^{-1} and 1000 min^{-1}, the coating layer completely separated from the Al carrier foil within 5 to 20 h. The time to a complete separation depended on the stirrer speed and the temperature, but it also depended, particularly, on the size of the cathode pieces, on the thickness of the coating, and on the quality of the adhesion between the coating and the Al substrate. Local damage to the coating, local lift-offs, or rough cutting edges (caused by mechanical comminution processes) strongly supported the detachment process. If the stirring was interrupted during the treatment, a weak development of gas bubbles could be observed (for an explanation, see Section 3.4), which evolved mainly from the cutting edges and from the cracks in the coating.

3.2. *The Release of Li and the Increase of the pH*

When a cathode comes into contact with water, the pH value rises sharply in the first few minutes. The pH value of the deionized water of 5.8 to 6.5 rose to values between 9 and 11 in the course of 30–60 min. The rate and extent of the pH increase depended on the ratio of the mass of the cathode to the volume of water, on the size of the cathode pieces, and the degree of damage to the coating and the Al substrate as a result of the mechanical comminution.

Li was released from the cathode without a time delay as soon as the cathode came into contact with water. At room temperature, the amount of Li did not subsequently increase significantly over time (Figure 3a). The amount of Li released was small in relation to the total Li content in the coating and, as in the example shown in Figure 3b, it reached a Li leaching efficiency of almost 1%. The correlation between the amount and the rate of Li release and the NMC content in the coating, as shown in Figure 4a, was only given if cathode foil pieces of a defined size, with smooth-cut edges, and without visible damage to the coating were used.

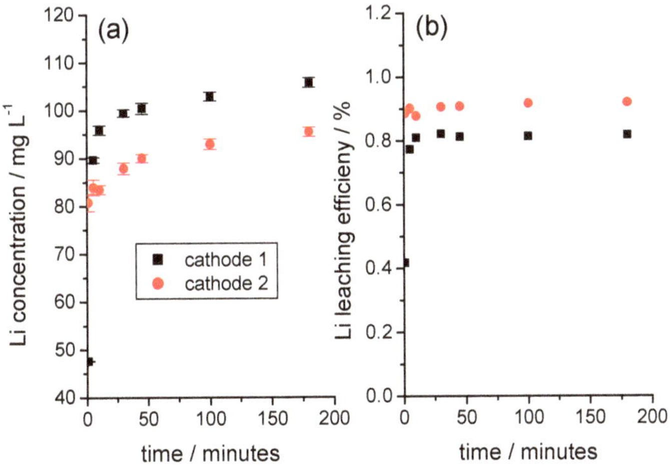

Figure 3. Initial period of the release of Li from two different cathode foils during immersion in water. (**a**) Solution concentration of Li as a function of time, (**b**) time dependence of the Li leaching efficiency calculated with respect to the total Li content in the sample (weight of cathode 1 sample: 1.1368 g; weight of cathode 2 sample: 1.1355 g).

A plausible model for the initial deintercalation of Li from NMC was recently discussed by Billy et al. [15]. A protonation of the surface of the oxidic NMC surface leads to the formation of –OH and $–OH_2^+$ species. Such a protolytic reaction at the solid–liquid interface is equivalent to an accumulation of a positive surface charge. This charge difference can be balanced by a partial internal oxidation of the transition metals and by the deintercalation of Li ions that are transferred into the aqueous solution. If this mechanism is valid for the initial contact between the NMC and an aqueous solution, then the initial increase in the pH is not caused by a generation of OH^- ions, but rather by a significantly reduced concentration of H^+ ions. Since the amount of H^+ ions attached to the NMC surface is finally limited, the amount of OH^- ions is limited too.

3.3. The Release of Sulfur

Figure 4 shows the time-dependent development of the solution concentrations for the element's Li and S, released from a sample of square pieces of cathode 2 in contact with water at 20 °C and 55 °C. To compare both experiments, the solution concentrations were normalized to the respective masses of the NMC in each studied cathode sample. Figure 4a shows that the initial normalized concentrations of Li at both temperatures are nearly identical. Over time, the normalized Li concentration increased only slightly at 20 °C; by contrast, the normalized Li concentration increased significantly at 55 °C, indicating that the NMC degraded significantly more at the elevated temperature.

The S concentrations (normalized to the mass of NMC in the samples) rose abruptly upon direct contact of the cathodes with water (Figure 4b) and reached similar normalized initial concentrations that remained almost unchanged over time. Surprisingly, the initial degradation of the NMC at 55 °C (Figure 4a) does not cause a further increase in the S concentration. The release of Li and S upon direct contact with water indicated that a slightly soluble Li compound was present on the surface of the NMC particles. Analysis of the solutions using ion chromatography showed that the S measured by ICP-OES was quantitatively present as sulfate ions. The molar ratio of released Li and S as sulfate $n(Li^+):n(SO_4^{2-})$ for the starting material at the first measuring point was 3.14:1, which suggested that the readily soluble Li compound was mainly lithium sulfate, Li_2SO_4, which went directly into solution upon contact with water and did not increase the pH value. Therefore, it was assumed that the release of Li beyond the stoichiometry of Li_2SO_4 was essentially caused by a degradation of the NMC.

The ion chromatographic analysis of the starting material also revealed, in addition to sulfate, very low levels of phosphate and acetate near the detection limit. It is known that NMC may also contain lithium oxide and lithium peroxides [15–17]. However, it was assumed that their potential contribution to the release of Li was very small, and thus they were neglected. Therefore, the increase in the pH must be attributed to the degradation of the NMC.

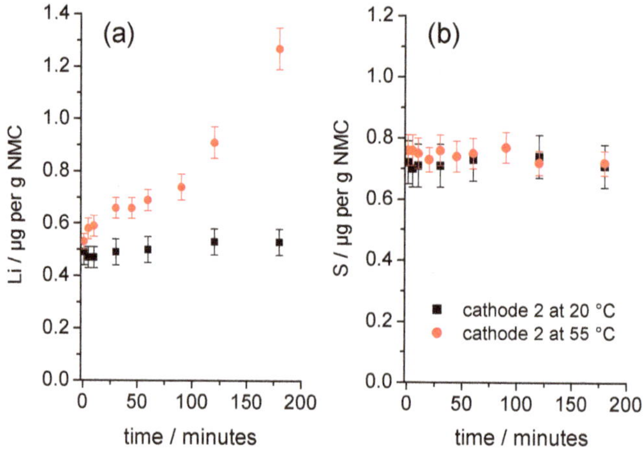

Figure 4. Time dependency of the release of (**a**) Li and (**b**) S from samples of cathode 2 immersed in water at 20 °C and 55 °C.

3.4. Dissolution and Precipitation of Al

With an increasing pH, the Al of the substrate (Figure 5a) began to dissolve, resulting in local bubble formation, preferably in places with damaged coatings and at the edges of the Al foil. According to Equation (1), from a pH value above 6, the easily soluble tetrahydroxoaluminate complex $[Al(OH)_4(H_2O)_2]^-$ is formed, as in Reference [18]:

$$2Al + 2OH^- + 10H_2O \rightarrow 2[Al(OH)_4(H_2O)_2]^- + 3H_2 \quad (1)$$

This complex can be converted to the hexahydroxoaluminate complex $[Al(OH)_6]^{3-}$ (Equation (2)) with an increasing pH:

$$[Al(OH)_4(H_2O)_2]^- \underset{+H^+}{\overset{-H^+}{\rightleftharpoons}} [Al(OH)_5(H_2O)]^{2-} \underset{+H^+}{\overset{-H^+}{\rightleftharpoons}} [Al(OH)_6]^{3-} \quad (2)$$

As long as the solution is in an alkaline environment, the concentration of the dissolved Al increases with the reaction time (Figure 5a). At a certain point, the Al concentration drops significantly, and the formation of a white precipitate becomes visible. The starting precipitation is accompanied by a decrease in the pH value, as illustrated in Figure 5b, using the example of a low-aluminum, mechanically shredded cathode foil fraction. With the ongoing dissolution of Al according to Equation (1), the concentration of the hydroxide ions is exhausted and the hydroxoaluminate complexes are no longer stabilized, which is equivalent to a reverse of Equation (2). Finally, the tetrahydroxoaluminate complex reacts with the Al, and the dissolved Al is almost completely precipitated as aluminum hydroxide, $Al(OH)_3$, according to Equation (3):

$$Al + [Al(OH)_4]^- + 2H_2O \rightarrow 2Al(OH)_3 + H_2 \quad (3)$$

In some series of measurements, the pH values > 11 were measured, which might lead to a precipitation of lithium aluminum hydroxide, Li[Al(OH)$_4$], according to Equation (4) [19].

$$Li^+ + [Al(OH)_4]^- \rightarrow Li[Al(OH)_4] \quad (4)$$

Reproducible, time-dependent curves for the solution concentrations of Li (Figure 3a) and Al (Figure 5a) were only obtained for defined samples of cathode foils. In these cases, the concentrations of Li and Al were in relation to each other, as long as the precipitation of Al(OH)$_3$ had not occurred. Mechanically shredded cathodes with an undefined sample condition, as shown in Figure 4b, led to a significant decrease in the reproducibility of the test results.

In contrast to the dissolution of lithium, sulfate, and aluminum, only very low concentrations of Ni, Mn, and Co could be detected in solution.

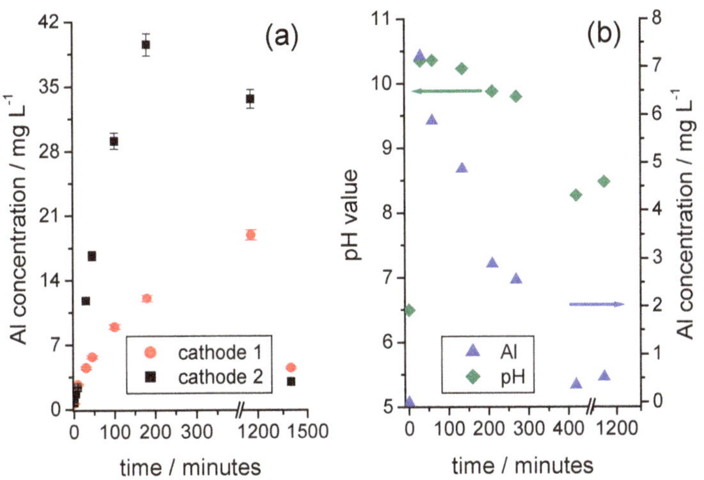

Figure 5. (**a**) Time-dependent development of the solution concentration of Al when the cathode foils are placed in water (sample weight of foil 1: 1.1368 g; sample weight of foil 2: 1.1355 g); (**b**) Time-dependent development of the solution concentration of Al and the pH value for a sample of a mechanically shredded cathode fraction in contact with water (sample weight: 3.0053 g).

Al(OH)$_3$ precipitates on the NMC particles. The SEM image in Figure 6a shows a secondary particle coated with a diffuse, cloudy-like coating that covers the particles either completely (position 1) or partially (position 2). Only in the middle area (position 3) no Al(OH)$_3$ coating appears to be present. The EDX element mapping for Al in Figure 6b confirmed the inhomogeneous distribution of Al across the measured particle. Compared to the high Al intensity at position 1, and a lower intensity of Al at position 2, almost no indications were found for a precipitation of Al(OH)$_3$ at position 3. Figure 6c shows the Raman spectra of the three selected regions, which proved that at each position of the NMC particle, Al(OH)$_3$ was detectable by the signal at 1052 cm^{-1} (the Al–O bending mode [20]), even at position 3, which was apparently assumed to be free of Al(OH)$_3$ according to the EDX mapping in Figure 6b. Thus, the Al(OH)$_3$ was deposited all over the entire surface in more or less thick layers on the NMC particles.

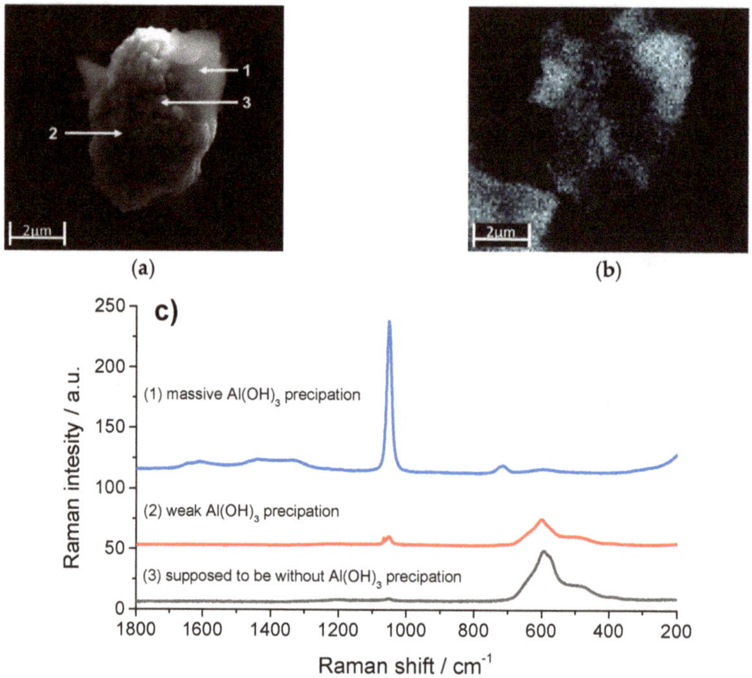

Figure 6. (**a**) SEM image of a NMC particle partially covered with $Al(OH)_3$. (**b**) The corresponding SEM/EDX mapping of the element distributions of Al (grey: High Al intensity). (**c**) Comparison of the Raman spectra from different regions of the particle: (1) Visible precipitation of $Al(OH)_3$, (2) region with apparently low $Al(OH)_3$ coverage, (3) region without SEM/EDX detectable $Al(OH)_3$ coverage.

3.5. Use of Buffer Solutions for Separation of The Cathode Coating

The precipitation of $Al(OH)_3$ on the particles can lead to a passivation, with a downgrading of the electrochemical properties that might void any re-use of the recovered NMC. One option to avoid the precipitation of $Al(OH)_3$ on the NMC particles is to perform the separation of the coating from the Al substrate layer in the acidic pH range. However, this will also lead to the attack and degradation of the NMC [15]. The second option, chosen in this paper, was the separation of the coating from the Al substrate in the alkaline pH range, where a sufficiently high hydroxide ion concentration was provided to convert the dissolved Al quantitatively into the water-soluble form of a hydroxoaluminate complex. KH_2PO_4/K_2HPO_4 buffer solutions, with a high buffer capacity to maintain a constant pH value over the entire leaching period, were investigated for their potential to separate the coating from the Al substrate and to avoid a precipitation and further chemical degradation of the NMC.

To ensure a constant pH value over the treatment time, the buffer capacity (β) was estimated using Equation (5):

$$\beta = \ln 10 \times \frac{c_{H_2PO_4^-} \times c_{HPO_4^{2-}}}{c_{H_2PO_4^-} + c_{HPO_4^{2-}}} \tag{5}$$

where $c_{H_2PO_4^-}$ and $c_{HPO_4^{2-}}$ are the concentrations of dihydrogen phosphate and hydrogen phosphate present in the buffer mixture. The ratio of these concentrations was previously calculated from the Henderson–Hasselbalch equation (Equation (6)):

$$\frac{c_{HPO_4^{2-}}}{c_{H_2PO_4^-}} = 10^{pH-pKs} \tag{6}$$

for the desired pH value of the leaching solution. Under the assumption that the Li removal does not exceed a value of 5% of the Li content in the used NMC material, and in the boundary case $c_{Li^+} = c_{OH^-}$, the maximum tolerable amount of NMC per a given buffer volume can be calculated for different desired pH values—i.e., for different initial concentrations $c_{H_2PO_4^-} = c_{HPO_4^{2-}}$.

Since the dissolution rate of the NMC increases rapidly with an increasing acid concentration [15], only one KH_2PO_4/K_2HPO_4 buffer solution in the acidic range (pH of 6) was investigated. In parallel, a test was carried out under the same conditions with deionized water as a reference. Unlike in the buffered solutions, the pH value of the deionized water changed over the entire course of the test, as shown in Figure 5b.

As Figure 7a shows, Li quickly dissolves at every investigated pH value. The contents achieved after approximately 20 min were already close to the maximum concentrations after 1400 min. The Li contents in the buffer solutions were comparably high and were above the values achieved in the aqueous solution, regardless of the pH value. Figure 7b shows the behavior of Al. In an aqueous solution, the Al content initially increases sharply and then decreases again due to the precipitation of $Al(OH)_3$ (which is also illustrated in Figure 5b). In the buffered solutions, however, no precipitation could be observed. In the pH range of 6 to 8, the Al concentration increased only slightly with the pH value, because the buffer solutions compensated for the initially high pH values between 9 and 11 (Figure 5b). Only at a pH value of 9 did a significant attack on the substrate occur, which coincided with a stronger evolution of bubbles. The Al concentration increased continuously and reached a higher solution concentration than in the parallel experiment with deionized water, because of the formation of the water-soluble hydroxoaluminate complexes in excess of the hydroxide ions.

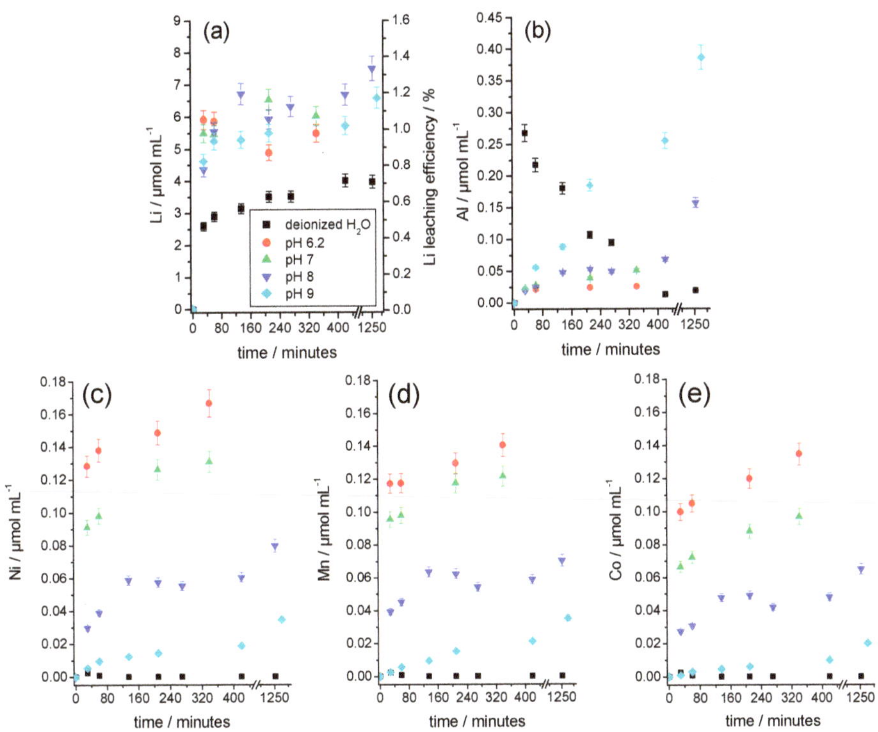

Figure 7. Time-dependent development of the elemental concentrations and the Li leaching efficiency in water and buffered alkaline solutions for the leaching of mechanically shredded cathodes. (**a**) Li and Li leaching efficiency, (**b**) Al, (**c**) Ni, (**d**) Mn, and (**e**) Co. Sample weights for each experiment 3.0 g.

Figure 7c–e shows the time dependence of the solution concentrations of Ni, Mn, and Co. These elements were only slightly dissolved in the buffered solution, whereby their solution concentrations increased continuously with a decreasing pH, as expected. A precipitation of Mn, Ni, and Co as divalent hydroxides was not plausible because their solubility constants (2.1×10^{-13} $mol^3{\cdot}L^{-3}$, 5.5 $\times 10^{-16}$ $mol^3{\cdot}L^{-3}$, and 1.1×10^{-15} $mol^3{\cdot}L^{-3}$) were significantly higher than those of the much less soluble $Al(OH)_3$ with 6×10^{-33} $mol^4{\cdot}L^{-4}$ [21].

The examination of the treated NMC samples by SEM-EDX and Raman spectroscopy showed no indications for a deposition of $Al(OH)_3$. Furthermore, there were no indications of degradation detected by X-ray powder diffraction, such as the formation of a phase, with a birnessite-type layered structure as observed by Billy et al. [15]. The chemical analysis of the treated samples gave stoichiometries that were identical to the initial stoichiometry within the range of uncertainty (c.f. Section 2).

3.6. Secondary Particle Disintegration

One criterion for the reuse of the recovered NMC is that the particle size and morphology should remain substantially unchanged by the treatment. So far, the removal of the coating in the buffered alkaline media did not cause any detectable change in the size or morphology of the NMC secondary particles. As shown before, the majority of the leached Li comes from the dissolution of Li_2SO_4, which presumably covers the outer surface of the NMC particles, and not by the Li^+ vs. H^+ exchange, nor by a significant degradation of the NMC particles.

However, a relationship between Li leaching efficiency and secondary particle disintegration was found. The NMC particles were treated with an aqueous citric acid solution (concentration 10^{-2} mol L^{-1}, pH 2.93, 25 °C) to accelerate the dissolution process. The SEM images in Figure 8 show the change of the particles with increasing treatment duration. After 30 min and a Li leaching of 0.5%, no change was visible (Figure 8). After 300 min and a Li removal of 1.6%, the first changes in the NMC particles were visible, as the particle size was reduced, and many single primary particles or small particles (which consist of a group of a few single primary particles) increased. This evolution continued with a longer treatment time, such that after 1815 min and a Li removal of 2.2%, a large decay of the secondary particles occurred. At the same time, Ni, Mn, and Co also dissolved as a result of a chemical attack on the NMC particles. In the experiments with the buffered alkaline solutions, a maximum Li leaching of approximately 1.3% was achieved at pH 8 after 1250 min (Figure 7a). This was only slightly below the value that clearly marked the disintegration of the secondary particles during treatment with citric acid. Even if considerably longer leaching times in alkaline leaching solutions are required to achieve a Li reduction of approximately 1.6%, the experimental duration is a decisive process parameter for the recovery of the NMC. In a technical implementation, it has to be considered that the reaction conditions are far less controllable than in a laboratory experiment, and that residence times in the associated technical facilities and conveyor sections must be taken into account, which can significantly increase the contact time between the NMC and the basic medium.

3.7. Raman Spectroscopy Studies

Raman spectroscopy applied to the NMC-type material with a $D_{3d}{}^5$ space group can probe the Raman active M–O symmetrical stretching vibrations, the A_{1g} mode, and the O–M–O bending vibrations, the E_g mode (M = Ni, Mn, Co) [2,22,23].

(**a**) (**b**) (**c**)

Figure 8. SEM images of the NMC particles (starting material after treatment with citric acid (pH 2.93, 25 °C, concentration 10^{-2} mol L^{-1}) for (**a**) 30 min, (**b**) 300 min, and (**c**) 1815 min. The corresponding Li leaching efficiencies were 0.5% for a, 1.6% for b, and 2.2% for c.

In this study, Raman microscopy was used to detect changes to the NMC particles as a result of their treatment with water and buffered alkaline solutions. The deconvolution and assignment of the measured Raman spectra, shown exemplarily in Figure 9a, was made according to Zhang et al. [2]. Figure 9b shows a plot of the normalized peak areas for the deconvoluted modes of NMC, the particles before and after treatment with water for 180 min in the absence of Al. The Raman measurements were made at arbitrarily chosen spots on the sample of water-treated NMC. The mean peak areas for a given vibration mode were represented by the heights of the columns, and the error bars represented the spread of the individual area values. Figure 9b represents the scatter of the individual Raman spectra originating from the local inhomogeneities in the starting material and reproduced in the treated NMC material. It was obvious that a reliable statement about a possible localized degradation of the NMC particles could not be derived.

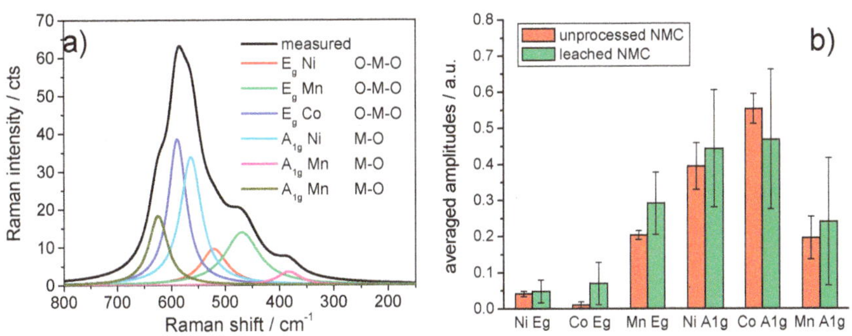

Figure 9. (**a**) Deconvolution of an NMC Raman spectrum and assignment of the subspectra according to Reference [2]. (**b**) Plot of the normalized subspectra amplitudes extracted from the deconvolution of Raman spectra of the NMC starting material and the NMC treated for 180 min with water. The column heights indicate the mean of the normalized areas, and the bars represent the width of scatter of the individual results.

The extent of local inhomogeneities underlined the normalized Raman spectra recorded from the individual NMC particles of different diameters of the starting material in Figure 10a, and after water treatment in Figure 10b. Each particle, regardless of size and even at an identical diameter, seemed to exhibit its own Raman spectrum. These studies showed the challenge in identifying suitable recovered NMC for further re-use in Li-ion batteries by Raman spectroscopy. Nevertheless, these results are the starting point for further studies.

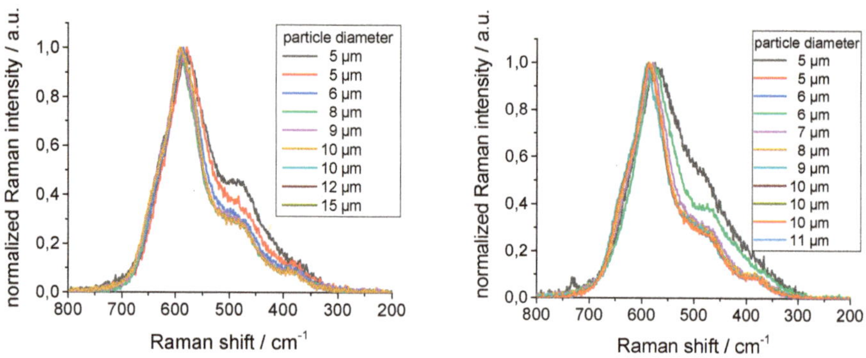

(**a**) (**b**)

Figure 10. Raman spectra of the individual NMC secondary particles of different diameters (**a**) of the starting material and (**b**) after water treatment for 180 min.

4. Discussion

If NMC comes into contact with water, a protonation of the NMC surface takes place, which leads to M–OH_2^+ species at the surface. The enrichment of the positive charge at the surface requires a charge balance, which leads to the deintercalation of Li^+. In practice, this refers to an exchange of Li^+ vs. H^+, which reduces the H^+ ion concentration in solution and thus increases the pH value and leads to the dissolution of Al. However, not all the NMC is available for protonation due to the embedding in the PVDF polymer. Furthermore, diffusion processes, partial under-etching of the coating, and cracks and deformations can influence the protonation and deintercalation of Li, as well as the dissolution of Al. This means that the defined correlations between the solution contents of Li^+ and Al^{3+}, as well as the time-dependent leaching behavior, can hardly be reproduced to a high quality. If the Al is initially dissolved as a hydroxoalumina complex, the continued dissolution of the Al leads to a reduction in the hydroxide ion concentration, which finally leads to the precipitation of $Al(OH)_3$, and possibly $LiAl(OH)_4$. The insoluble $Al(OH)_3$ precipitates in several or a few thick layers on the NMC particles. This passivation is considered to downgrade the electrochemical performance of the recovered NMC particles and might prevent their re-use.

The precipitation of $Al(OH)_3$ on the NMC particles can be avoided if alkaline solutions with sufficient buffer capacity are used instead of water. Although an attack on the Al substrate film also occurs in the alkaline medium, the Al remains in solution in excess of the hydroxide ions, forming highly water-soluble hydroxoaluminate complexes. The alkaline medium leads to a partial attack on the NMC, which manifests itself in low solution concentrations of Ni, Mn, and Co. The compromise between the side reactions of dissolving Al and dissolving Ni, Mn, and Co is the choice of a pH value between 7 and 8. At the end of the experiments, the NMC particles are almost identical in size and morphology as the starting material used.

The amount of Li released in the experiments was small in relation to the total Li content in the studied cathode samples. Two parallel proceeding mechanisms seem to determine the release of Li: (i) The degradation of the NMC according to Billy et al. [15] and (ii) a release via the dissolution of traces of inorganic salts, in this study, presumably Li_2SO_4. In the alkaline buffered media, typical Li leaching efficiency values were in the range of 1%, and the maximum value obtained after the longest leaching time of 1250 min was found to be 1.3%. Although this value was low and was determined mainly by the highly soluble Li salts (Li_2SO_4), a higher removal of Li should be avoided. Investigations in the acid solution showed that a Li removal of approximately 1.6% disintegrates the NMC secondary particles, generating significantly smaller secondary particles, which are particles consisting of only a few primary particles, and even single primary particles. Thus, the salts present on the surface contribute significantly to the cohesion of the secondary particles. This demonstrates how sensitively the chemical treatment influences the particle size, and thus, the quality of the recovered NMC. Finally, Raman spectroscopy is typically used to characterize NMC and to evaluate the quality of the recovered NMC. However, a reasonable interpretation of the Raman spectra was limited because the NMC starting material, as well as the treated NMC, were inhomogeneous. The resulting Raman spectra of the starting material and the treated NMC exhibited such a scattered range that clear statements about the quality of the treated NMC cannot be made at the present time. Further efforts—in particular, electrochemical studies on the performance of the recovered NMC—are necessary to investigate this phenomenon.

5. Conclusions

This paper described the first step of an approach, designated as functional recycling, to recover NMC material from the cathodes of end-of-life Li-ion batteries or scrap from the production of cathodes, while preserving its chemical, physical, and morphological properties. The separation of the cathode coating from the Al substrate foil can be performed in a neutral or alkaline medium, with minimal mechanical treatment. However, several chemical side reactions and side effects, such as the release of Li, the formation of a passivation layer, or the disintegration of the secondary NMC particles, were identified to have a crucial effect on the quality of the recovered NMC material. Furthermore, this paper demonstrated the emerging challenges to the analytical methods and tools to detect the smallest changes in the material, to develop the most efficient and environmentally friendly route to recover NMC for re-use in Li-ion batteries.

Author Contributions: Conceptualization, J.D. and J.A.; methodology, J.D., A.R., T.S., and J.A.; leaching experiments, J.D., A.R., and T.S.; Raman and SEM-EDX investigation, T.S. and T.L.; formal analysis, T.S. and T.L.; supervision, J.A.; all authors contributed to the production of the manuscript.

Funding: This research was funded by Bundesministerium für Bildung und Forschung (BMBF) under grant number 16SBS007.

Conflicts of Interest: The authors declare no conflict of interest.

References

1. Blomgren, G.E. The Development and Future of Lithium Ion Batteries. *J. Electrochem. Soc.* **2017**, *164*, A5019–A5025. [CrossRef]
2. Zhang, X.; Mauger, A. Synthesis and characterization of $LiNi_{1/3}Mn_{1/3}Co_{1/3}O_2$ by wet-chemical method. *Electrochimica Acta* **2010**, *55*, 6440–6449. [CrossRef]
3. Study on the Review of the List of Critical Raw Materials-Final Report. EU Publications 2017. Available online: https://publications.europa.eu/en/publication-detail/-/publication/08fdab5f-9766-11e7-b92d-01aa75ed71a1/language-en (accessed on 20 November 2018).
4. Study on the Review of the List of Critical Raw Materials-Critical Raw Materials Factsheets. EU Publications 2017. Available online: https://publications.europa.eu/en/publication-detail/-/publication/7345e3e8-98fc-11e7-b92d-01aa75ed71a1/language-en/format-PDF/source-86374692 (accessed on 20 November 2018).
5. Zhang, P.; Yokoyama, T. Hydrometallurgical process for recovery of metal values from spent lithium-ion secondary batteries. *Hydrometallurgy* **1998**, *47*, 259–271. [CrossRef]
6. Lain, M.J. Recycling of lithium ion cells and batteries. *J. Power Sources* **2001**, *97–98*, 736–738. [CrossRef]
7. Chagnes, A.; Pospiech, B. A brief review on hydrometallurgical technologies for recycling spent lithium-ion batteries. *J. Chem. Technol. Biotechnol.* **2013**, *88*, 1191–1199. [CrossRef]
8. Gaines, L.; Dunn, J. *Lithium-Ion Batteries*, 1st ed.; Elsevier Science Ltd.: Oxford, England, 2014; pp. 483–508, ISBN 978-0-444-59513-3.
9. Vazzini, A. *Lithium-Ion Batteries*, 1st ed.; Elsevier Science Ltd.: Oxford, England, 2014; pp. 529–551, ISBN 978-0-444-59513-3.
10. Zeng, X.; Li, J. Recycling of Spent Lithium-Ion Battery: A Critical Review. *Crit. Rev. Environ. Sci. Technol.* **2014**, *44*, 1129–1165. [CrossRef]

11. Ekberg, C.; Petranikova, M. Lithium Batteries Recycling. In *Lithium Process Chemistry: Resources, Extraction, Batteries, and Recycling*, 2nd ed.; Chagnes, A., Swiatowska, J., Eds.; Elsevier: Amsterdam, Netherlands, 2014; pp. 233–267, ISBN 978-0-12-801417-2.
12. Sonoc, A.; Jeswiet, J. Opportunities to Improve Recycling of Automotive Lithium Ion Batteries. *Procedia CIRP* **2015**, *29*, 752–757. [CrossRef]
13. Zheng, R.; Wanf, W. A closed-loop process for recycling $LiNi_xCo_yMn_{(1-x-y)}O_2$ from mixed cathode materials of lithium-ion batteries. *Green Energy Environ.* **2017**, *2*, 42–50. [CrossRef]
14. Kwade, A.; Diekmann, J. *Recycling of Lithium-Ion Batteries-The LithoRec Way*, 1st ed.; Springer International Publishing AG: Cham, Switzerland, 2018; pp. 207–246, ISBN 978-3-319-70571-2.
15. Billy, E.; Joulié, M.; Laucournet, R.; Boulineau, A.; De Vito, E.; Meyer, D. Dissolution Mechanisms of $LiNi_{1/3}Mn_{1/3}Co_{1/3}O_2$ Positive Electrode Material from Lithium-Ion Batteries in Acid Solution. *ACS Appl. Mater. Interfaces* **2018**, *10*, 16424–16435. [CrossRef] [PubMed]
16. Sathiya, M.; Ramesha, K. High Performance $Li_2Ru_{1-y}Mn_yO_3$ ($0.2 \leq y \leq 0.8$) Cathode Materials for Rechargeable Lithium-Ion Batteries: Their Understanding. *Chem. Mater.* **2013**, *25*, 1121–1131. [CrossRef]
17. Shimoda, K.; Minato, T. Oxidation Behaviour of Lattice Oxygen in Li-rich Manganese-based Layered Oxide Studied by Hard X-ray Photoelectron Spectroscopy. *J. Mater. Chem. A* **2016**, *4*, 5909–5916. [CrossRef]
18. Hollemann, A.F.; Wiberg, E.; Wiberg, N. *Lehrbuch der Anorganischen Chemie*, 101st ed.; Walter de Gruyter & Co.: Berlin, Germany, 1995; pp. 1077–1080, ISBN 3-11-012641-9.
19. Champenois, J.B.; Mesbah, A. $LiAl_2(OH)_6OH{\cdot}2H_2O$ solubility product and dihydrogen radiolytic production rate under g-irradiation. *J. Nucl. Mater.* **2018**, *508*, 92–99. [CrossRef]
20. Cherepy, N.J.; Shen, T.H. Characterization of an effective cleaning procedure for aluminum alloys: Surface enhanced Raman spectroscopy and zeta potential analysis. *J. Colloid Interface Sci.* **2005**, *282*, 80–86. [CrossRef] [PubMed]
21. Küster, F.W.; Thiel, A. *Rechentafeln für die Chemische Analytik*, 105th ed.; De Gruyter: Berlin, Germany, 2002; pp. 224–225, ISBN 978-3110175660.
22. Baddour-Hadjean, R.; Pereira-Ramos, J.-P. Raman Microspectrometry Applied to the Study of Electrode Materials for Lithium Batteries. *Chem. Rev.* **2010**, *110*, 1278–1319. [CrossRef] [PubMed]
23. Ben-Kamel, K.; Amdouni, N.; Mauger, A.; Julien, C.M. Study of the local structure of $LiNi_{0.33+\delta}Mn_{0.33+\delta}Co_{0.33-2\delta}O_2$ ($0.025 \leq \delta \leq 0.075$) oxides. *J. Alloys Compd.* **2012**, *528*, 91–98. [CrossRef]

nanomaterials

MDPI

Article

ToF-SIMS 3D Analysis of Thin Films Deposited in High Aspect Ratio Structures via Atomic Layer Deposition and Chemical Vapor Deposition

Alireza M. Kia [1,*], Nora Haufe [1], Sajjad Esmaeili [1], Clemens Mart [1], Mikko Utriainen [2], Riikka L. Puurunen [2,3] and Wenke Weinreich [1]

1 Fraunhofer Institute for Photonic Microsystems, 01099 Dresden, Germany
2 VTT Technical Research Centre of Finland Ltd., 02044 Espoo, Finland
3 School of Chemical Engineering, Aalto University, 02150 Espoo, Finland
* Correspondence: alireza.mohammadian.kia@ipms.fraunhofer.de; Tel.: +49-351-2607-3207

Received: 30 April 2019; Accepted: 16 July 2019; Published: 19 July 2019

Abstract: For the analysis of thin films, with high aspect ratio (HAR) structures, time-of-flight secondary ion mass spectrometry (ToF-SIMS) overcomes several challenges in comparison to other frequently used techniques such as electron microscopy. The research presented herein focuses on two different kinds of HAR structures that represent different semiconductor technologies. In the first study, ToF-SIMS is used to illustrate cobalt seed layer corrosion by the copper electrolyte within the large through-silicon-vias (TSVs) before and after copper electroplating. However, due to the sample's surface topography, ToF-SIMS analysis proved to be difficult due to the geometrical shadowing effects. Henceforth, in the second study, we introduce a new test platform to eliminate the difficulties with the HAR structures, and again, use ToF-SIMS for elemental analysis. We use data image slicing of 3D ToF-SIMS analysis combined with lateral HAR test chips (PillarHall™) to study the uniformity of silicon dopant concentration in atomic layer deposited (ALD) HfO_2 thin films.

Keywords: ToF-SIMS 3D imaging; compositional depth profiling; high aspect ratio (HAR) structures; silicon doped hafnium oxide (HSO) ALD deposition; lateral high aspect ratio (LHAR); ToF-SIMS analysis

---

1. Introduction

Thin film deposition techniques such as atomic layer deposition (ALD), for three-dimensional microscopic structures and the development of large-scale integrated structures like trenches and cavities that are of importance, especially in the semiconductor industry, introduce a true challenge for thin film characterization. These 3D-micro-structured substrates have typically vertical high aspect ratio (VHAR) structures. The study and characterization of specific regions of interest like local defects, doping concentration, conformality, and interfaces, rely predominantly on cross-sectional sample preparation and characterization by electron microscopy or X-ray techniques. This approach faces the specific difficulties of single lamella preparation of each trench, low spatial resolution to characterize interfacial diffusion, or lack of detection of light elements in transmission electron microscopy (TEM) and cleavage plane inaccuracy for scanning electron microscopy (SEM).

To overcome the difficulties of the abovementioned characterization techniques, we used time-of-flight secondary ion mass spectrometry (ToF-SIMS) on HAR structure analysis. The SIMS technique is a powerful tool which provides information about a given material from both its surface, with an overall sensitivity of parts per million (ppm), and bulk, with a sensitivity of part per billion (ppb) [1,2]. The in-depth analysis of a thin film's chemical composition, as well as interfacial characterization, yields information about the elemental distribution in the range of a few monolayers

(1–3 monolayers). Furthermore, by using advanced liquid metal ion guns (LMIG), one can reach a lateral resolution of about 100 nm for the elemental detection (ToF-SIMS imaging) [3–5]. In addition to both remarkable in depth and lateral resolution, numerous parameters can affect the ToF-SIMS measurement technique that make data acquisition and interpretation challenging. For instance, the achievable spatial resolution in ToF-SIMS imaging is a function of the sample matrix, material concentrations, surface geometry, primary ion intensity, instrument transmission, and spot size of the primary ion beam [3,6,7].

In this paper, the first part of the study focuses on a non-copper-plated area at the bottom of large through-silicon-vias (TSVs). The TSV interconnects provide the shortest electrical pathway, lower power consumption, lower noise, smaller form factor, and yield better performance and more functionality in comparison with the conventional chip multi-layers stacking (CMLS) [8,9]. The most common TSV metallization stack is composed of a copper (Cu) diffusion barrier and a seed layer followed by the bottom-up being electroplated [10]. One of the main difficulties in the characterization of TSV metallization is the elemental analysis of thin barrier/seed layers through the depth of TSVs before and after Cu electroplating. The thin film characterization in the range of nanometers deep down into TSVs (within the range of hundred-micrometers) make it crucial to determine conformality, step coverage, purity and corrosion. With ToF-SIMS, we are able to check the corrosion under the influence of the Cu electrolyte [11] and map detailed elemental composition information onto the TSV wall.

In the second part, as an illustration, we choose to monitor the uniformity of silicon dopant concentration in the HfO_2 thin films deposition as atomic layer. The recent discovery of ferroelectric properties in this binary oxide material [12], commonly established as a dielectric for high-k metal gate technology (HKMG), handover a vast quantity of applications beyond the standard dielectric material. These applications span from non-volatile memories [13,14], steep slope devices [15] energy storage applications [16,17] and infrared sensors [18]. In this part of study, we focus on improving the analysis of the HAR structures by bypassing the obstacles of TSV characterization by ToF-SIMS and using LHAR test structures. ToF-SIMS in dual-beam mode (dynamic mode) is a well-established technique due to its high detection sensitivity for concentration ratio measurements of the dopant materials [19,20]. However, the analysis of these VHAR structures for quantifying dopant concentration is hardly possible. To solve this issue, we analyzed layer composition in a different system while being consistent with the results of the initial structures. In this system, instead of using the vertically oriented structures, lateral high aspect ratio (LHAR) [21–23] structures are used which make it possible to analyze them in the form of 2D-structures.

2. Experimental

2.1. Materials and Thin Films Preparation

2.1.1. TSV Metallization

The cobalt metal-organic chemical vapor deposition (Co-MOCVD) process enables manufacturing an adhesive and conformal thin film all the way through the TSVs, in comparison with the conventional non-conformal copper physical vapor deposition (Cu-PVD) process as a seed layer. However, the Co-MOCVD metallization mode requires achieving a proper seed layer with less organic contamination (e.g., carbon) which would intensify the Co corrosion under the Cu electrolyte influence. Due to the electroplating bath chemistry, and especially because of the Cu electrolyte component, the Co thin film is prone to corrosion during electrochemical deposition (ECD). The Cu ions content have great influence on Co corrosion. The redox reaction of Cu solidification is a fast corrosive factor. If the concentrations of the components are high enough in electrolyte, the cobalt corrosion reaction is faster than the Cu deposition. Due to this comparable faster dissolution of Co seed layer, there will most likely be non-copper-plated areas after the deposition process on the bottom of TSVs.

seed layer, there will most likely be non-copper-plated areas after the deposition process on the bottom of TSVs.

Figure 1 schematically shows the structure (Figure 1a) and the corroded bottom sidewalls after electroplating (Figure 1b). With respect to the Co layer, the TSVs' elemental and compositional depth profiles, especially on the bottom sidewall, were investigated before and after Cu-ECD. ToF-SIMS is used to measure the relative Co content in order to determine which combination of parameters yielded the highest and most stable amount of Co. The amount of Co ions result from ion gun sputtering through the depth of deposited films. Eventually, the comparative study with TEM was carried out to reveal the reliability of ToF-SIMS measurement.

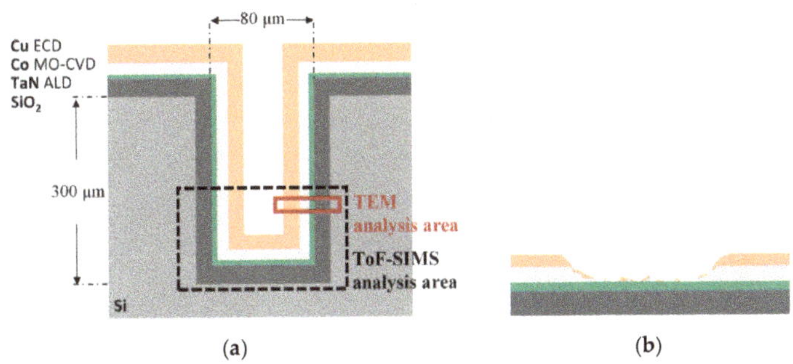

Figure 1. Schematic illustrations of (a) the ideal through-silicon-via (TSV) metallization process after Cu electrochemical deposition (ECD) and (b) the corroded cobalt-seed layer during Cu-ECD.

Silicon wafers, sized 300 mm, with dry etched wholes with an aspect ratio of 1:4 were chosen to perform the metallization process. All barrier film stack depositions were carried out without breaking the vacuum of the cluster tool, from Applied Materials (AMAT) Endura 2. Tantalum-Nitride (TaN) was employed as the Cu diffusion barrier layer which was fabricated by the ALD process for achieving an excellent conformality all the way through the TSVs. Afterwards, the Co thin film was fabricated at a temperature of 150 °C using the Hexacarbonyl (3,3-dimethyl-1-butyne) dicobalt (CCTBA) precursor at 5 Torr by MOCVD. Subsequently, the ECD process was accomplished on the coupon level in a lab-scale plating cell right after a pre-wetting treatment. The plating cell was set up with a connected cathode (the structured sample) and an anode (Cu metal plate) electrodes which were immersed inside a low-copper electrolyte. The electrolyte contained copper ions and sulfuric acid at concentrations of 4 g/L and 10 g/L, respectively.

2.1.2. HSO ALD Deposition

In this study, we used metal-organic Tetrakis (ethylmethylamido) hafnium (IV) (TEMAHf) and Tris (dimethylamino) silane (3DMAS) to form Si-doped HfO_2 (HSO). Superior conformality of ALD thin film deposition is a direct consequence of the inherent self-limiting reactions [24]. The main goal for ALD process development for deep trench capacitors is to find precursors which allow a conformal deposition in HAR structures and show the same material composition along the trench sidewall. To achieve that, it is necessary to study the precursors behavior in deep trench structures. Due to the different partial pressures, molecular size and molar masses of different precursors, the gas diffusion behavior into the trenches will be different for processes using more than one precursor. Therefore, varying dopant concentration levels may occur throughout the depth of trenches.

The absence of a simple and readily available 3D structure for elemental analysis of thin films produced by ALD led us to use a different system with the capability to analyze 3D HAR structures in the form of a 2D-planar surface. To be able to perform ToF-SIMS to optimize the deposition process, an LHAR structure is used, which is depicted schematically in Figure 2a. The microscopic LHAR structures were fabricated in chips on 150 mm silicon wafers using standard surface micromachining techniques. The chips contain multiple lateral cavities processed on top of single-crystal silicon with a polysilicon membrane roof sustained by polysilicon pillars [22]. The pillars provide a defined

Nanomaterials 2019, 9, x FOR PEER REVIEW 4 of 14

micromachining techniques. The chips contain multiple lateral cavities processed on top of single-crystal silicon with a polysilicon membrane roof sustained by polysilicon pillars [22]. The pillars provide a defined geometry with a nominal gap height of 500 nm. One chip contains LHAR structures (cavities) with different membrane lengths (L) from 1 µm to 5 mm (AR range 2:1–10,000:1), each with a single-crystal silicon area in front with defined width for easy identification (W = 100 µm, 90 µm, etc.). The roof of this LHAR test structure can be removed using adhesive tape and the deposited material can be assessed directly, as shown in Figure 2b. For thin film analysis with ToF-SIMS, we defined the analysis area at the 5 mm length membrane (W = 100 µm).

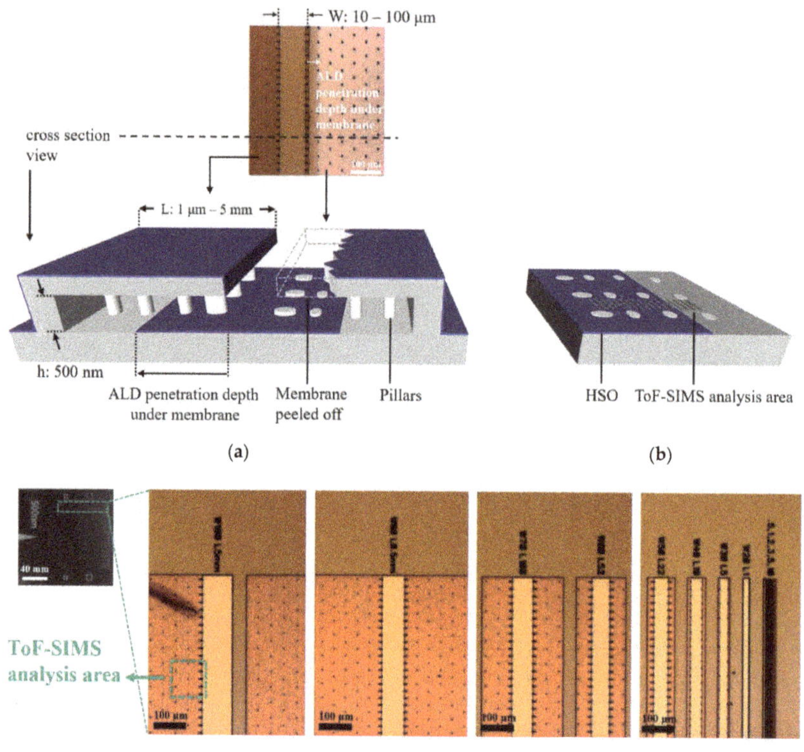

Figure 2. Schematic illustration of (a) the PillarHall™ test structure (bottom) that is used in the atomic layer deposited (ALD) process, and the corresponding top-view optical microscopic image (top) and (b) the test structure after membrane removal. The diffusion depth of the ALD process can be examined directly by time-of-flight secondary ion mass spectrometry (ToF-SIMS). Color code: Gray = silicon substrate, Blue = deposited material with ALD, Dark gray rectangles = ToF-SIMS region of interests (ROIs). h: cavity height, W: opening width, L: lateral length of membrane (images dimension are not to scale). In (c), the uncoated test chip and the optical microscope images of the corresponding top-view from different cavities with the membrane is shown.

The HSO material was processed at a temperature of 280 °C using a thermal ALD process at a Jusung Eureka 3000 (Tokyo, Japan). The TEMAHf and O_3 were used as a metalorganic precursor and oxidizer respectively for HfO_2 deposition. For SiO_2 doping, 3-DMAS/O_3 cycles were inserted to achieve a controlled ratio of dopant and HfO_2. During all processes, Ar gas was used for purging.

2.2. Analysis Tool Setups

The SIMS analysis data were acquired using a TOF-SIMS 300R (IONTOF GmbH, Münster, Germany) in order to analyze both overall film growth and doping levels inside the structures. The tool

Nanomaterials 2019, 9, 1035

2.2. Analysis Tool Setups

The SIMS analysis data were acquired using a TOF-SIMS 300R (IONTOF GmbH, Münster, Germany) in order to analyze both overall film growth and doping levels inside the structures. The tool is equipped with a bismuth (Bi) liquid metal primary ion gun (LMIG). The primary ion source was used with a short pulse at the voltage of 25 kV (anode) at 45°. The beam was chopped in 20 ns and focused to about 300 nm in diameter. The ion beam current was ~13 nA (DC mode). The secondary ions were collected with an extraction lens (biased at −40 V), traveled through the reflection flight tube and identified with a micro-channel plate (MCP) detector. A dose density of primary ions for each measurement was about 6.5×10^{-13} ions/cm². The vacuum pressure in the main chamber was kept at about 1.0×10^{-8} mbar.

A mass spectrum for each pixel was collected by raster-scanning of the primary ion beam to obtain a mass spectrometry data image across the sample area of interest (150 × 150 µm² using TSVs and 102 × 102 µm² using LHAR test chips). An electron impact (EI) gas ion source of oxygen (O_2^+) was used as a second ion beam source to remove material for profiling electropositive elements in positive SIMS. The energy of the sputter gun was set to 1 kV to concede the image profile of interesting fragments in depth. The average beam current was 0.23 µA. The initial velocity of secondary ions was provided by an accelerator into the drift tube with a potential of 2 kV. This velocity distribution was implemented for the mass separation by flight time analysis. A mass resolution of 8300 (m/Δm) at 28 amu can be accomplished with the flow time transformed into atomic mass units (amu). For the image analysis, data were acquired from 400 × 400 µm² by an ion beam raster of 256 × 256 pixels in a saw-tooth mode. Image analysis is conducted in SurfaceLab v6.3 (developed at IONTOF GmbH).

For step coverage analysis we used SEM by Hitachi S4800 (Hitachi, Japan) with an electron beam source energy at 10 kV using a back-scattered detector. For higher lateral resolution on the TSV cross-section, TEM observation was carried out for the structural analysis in localized areas at the bottom of the TSV sidewalls using FEI Tecnai F20 (Hillsboro, OR, USA). The acceleration energy of the electron beam was 200 keV. The process of lamella preparation for TEM observation is explained in Appendix A.

3. Results and Discussion

3.1. Vertical High Aspect Ratio Structures

As mentioned earlier, one of the problems with characterizing the elemental distribution in TSVs is the geometry of the structures in comparison with planar samples. The surface topography can cause undesirable artifacts in the SIMS image and spectra profile which significantly inhibits interpretation and quantification of data. As a consequence, surface evaluation and depth analysis with microscale topography are a considerable challenge due to analytical instrumental limitation. To study the inside of TSVs, the samples were cut and mounted on the sample holder in such a way that the cross sectioned TSV sidewalls face the analyzer. Figure 3 demonstrates the analysis area and sputter zone in the cross-sectional view.

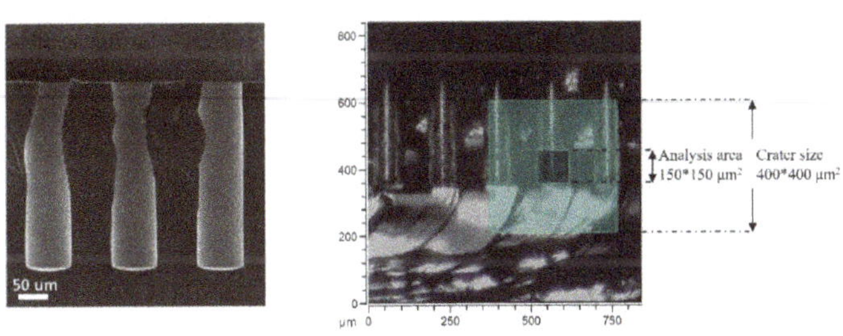

(a) (b)

Figure 3. Cross-sectional images showing (a) an SEM micrograph of TSVs and (b) a camera snapshot of a mounted sample (without Cu plating) showing the side wall with the crater size and analysis area, as highlighted in the image, during the ToF-SIMS measurement.

Figure 3. Cross-sectional images showing (**a**) an SEM micrograph of TSVs and (**b**) a camera snapshot of a mounted sample (without Cu plating) showing the side wall with the crater size and analysis area, as highlighted in the image, during the ToF-SIMS measurement.

A 3D reconstruction of the ToF-SIMS data was used to map the elemental distribution. The resulting image intensity is influenced by the effect of sample surface orientation or surface topography of the analysis area on the trajectory of the primary ion beam. Sputter yields not only depend on the primary ion energy but also the angle of incidence [2,25,26]. The angle of incidence changes depending on of the surface topography [3]. This induces changes in the energy distribution of the emitted secondary ions, the sputtering yield, and the angular distribution of ejected fragments (based on the emission angle from the surface normal) [27,28]. This artifact can cause falsified variations in image intensities (see Figure 4).

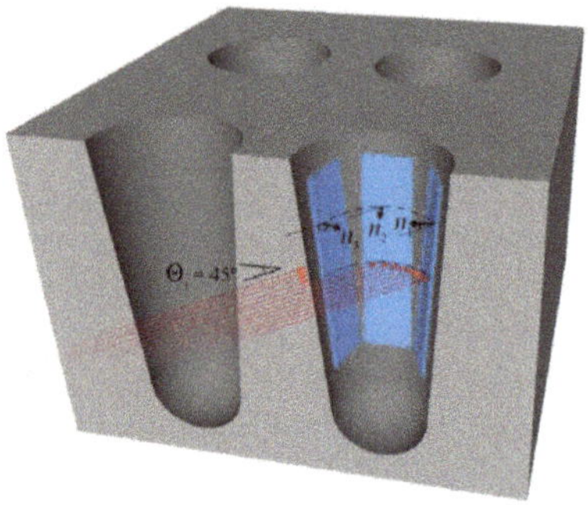

Figure 4. Schematic representation of a TSVs cross-section. Primary ion beams strike the sample's top surface at an incident angle of 45° (red arrows). The n_1, n_2, and n_3 denote the axes normal to each surface (blue colored planes). Each of these surfaces depict the relative angle between the incident ion beams and the inner surface of the TSV.

The acceptance angle of the primary ions (red arrows) with respect to the sample normal of the left inner surface (n_3), depicted in Figure 4, is limited in comparison to the middle and right inner surface (n_2, n_1). Consequently, the detection of secondary ions resulting from the right inner surface have a higher probability to reach the mass analyzer. Whereas, the secondary ions from the left inner surface can hardly reach the analyzer. The image profile of a 25 nm thick film of Co in the TSV sputtered using a 1 kV oxygen-ion beam, is shown in Figure 5a. Cobalt is visible as $^{59}Co^+$ (m/z = 58.93) throughout the TSV. Figure 6a, shows that Co is completely corroded after electroplating. As illustrated in Figures 5a and 6a, the same constructed layout, as in Figure 4, is built from ToF-SIMS analysis. The samples' orientations, with respect to the primary ion beams, and the analyzer perfectly agree with the schematic representation in Figure 4.

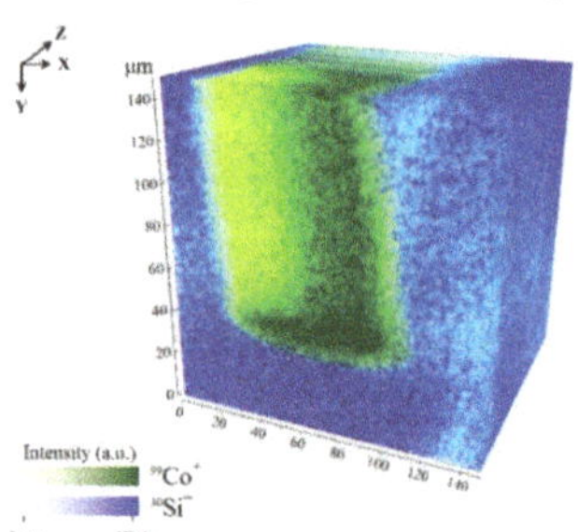

(**a**)

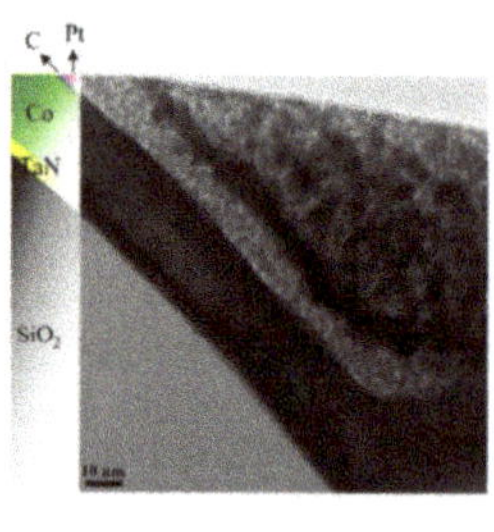

(**b**)

Figure 5. Images shown here are of the TSV before Cu electroplating. In (**a**) the 3D ToF-SIMS map of Si and Co signal from the bottom is shown and in (**b**) a TEM image from the bottom sidewall of the TSV showing the TaN barrier the cobalt film and the preparation fill.

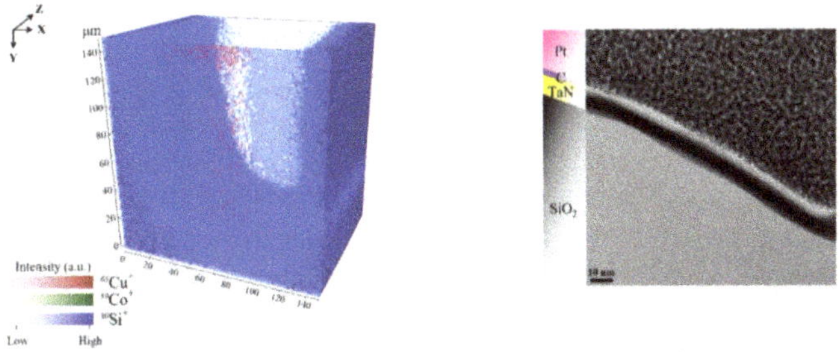

(a) (b)

Figure 6. Images shown here are of the TSV after Cu electroplating. In (**a**) the 3D ToF-SIMS map of Si, Co and Cu signals from the bottom and in (**b**) a TEM image from the bottom sidewall of the TSV is shown.

In Figures 5a and 6a, the geometrical influence on 3D imaging is pronounced for the trench sidewalls. A higher contrast is observed for Co and Cu of the sidewalls which face the primary ion gun. This confirms the influence of surface topography on SIMS 3D image reconstruction. In most SIMS instruments, the incident angle of the primary ions (viewing direction) creates an angle between 30° and 60°. As a result, the shadowing effects of the surface geometry are unavoidable [3]. Although, different approaches are suggested by Lee et al. such as sample alignment, changing extraction voltage, using cluster ion beam, and adjusting extraction delay to reduce the shadow effect [29]. However, this effect could not be eliminated entirely.

In addition, we compare the ToF-SIMS results with TEM micrographs which support the findings of ToF-SIMS shown in Figures 5b and 6b. Both TEM lamellas are lifted out from the bottom of the TSVs. The Pt and C that are marked in the figures are used as the protection layers in the process of lamella preparation for TEM to protect the top surface from unwanted surface damage by FIB milling and amorphization with high energetic ions. Comparing the results from TEM with ToF-SIMS, 3D images determined the Co corrosion at the bottom of TSV. The TEM image in Figure 5b shows the 25 nm thick Co seed layer on top of the TaN, which matches with ToF-SIMS analysis results. After Cu electroplating, it is not possible to detect Co species at the bottom of the TSV sidewall. Besides, Cu signal intensity is very low, which supports the idea of complete corrosion at the bottom of the TSV. To check the precision of this observation, a new lamella was prepared from the sample after electroplating. The results in Figure 6b support the assumption of Co thin film corrosion and failure on Cu deposition.

3.2. Lateral High Aspect Ratio Structures

One type of ferroelectric based non-volatile memory is the 1T-1C ferroelectric random-access memory (FRAM) where the memory state is stored in a ferroelectric capacitor. The readout current through the access transistor becomes proportional to the ferroelectric remnant polarization. In order to improve the sense current, a common 3D-like structure is utilized where the area factor enhances the total polarization charge [30]. Various dopants influence the stabilization of the ferroelectric phase in hafnia thin films, for example, silicon (Si) [17,31,32], and such influence becomes more decisive in deep trench structures while the dopants concentration may change inside the HAR structure. Remarkably, it is more critical in low-doped materials, where such a low Si content is necessary, while a small deviation results in strong change in the stabilized phase [33]. As compared to planar film deposition, effects on the layer composition and crystal structure are expected to be different for HAR structures. Therefore, a technique to quantify the Si concentration in ferroelectric films deposited in 3D structures is crucial for optimizing the thin film deposition process of FRAM applications.

structures. Therefore, a technique to quantify the Si concentration in ferroelectric films deposited in 3D structures is crucial for optimizing the thin film deposition process of FRAM applications.

Due to the small dimensions of 3D structures it becomes crucial to reach an attainable lateral resolution of different analytical techniques. In ToF-SIMS 3D imaging, the dimension of collision cascade and surface damage define the limitation that one could reach in the lateral resolution. Consequently, the useful lateral resolution of SIMS imaging relies on the primary ion beam focus and ionization efficiency of the surface analysis [34]. In order to reach the maximal lateral resolution, it is a compromise between acquisition time (number of ions per shot), mass resolution (pulse length) and lateral resolution (spot size) is unavoidable. To succeed in high mass resolution, short pulses in the order of <1 ns are required. Producing short pulses with higher energy cost deterioration of the focus in the range of 3–10 µm. In contrast, to achieve a lateral resolution in the range of 200–300 nm, it is required to increase the pulse length larger than 150 ns whilst decreasing the mass resolution. In spite of such a trade-off, it is also possible to simultaneously obtain high spatial and high mass resolution as described by Vanbellingen et al. by applying delay extraction of secondary ions [35].

Figure 7a,b show SEM micrographs of VHAR structures with 20 nm of HSO deposited on them. The non-planar surface geometry of the collected samples' cross sections influences the secondary ion yield and restricts the number of ions that can be created out of the sputtered area. In addition, the surface topography has an effect on the ultimate mass resolution, which arises from the slight differences in the extraction field [36]. In low-intensity imaging, many small features are affected by artifacts of counting statistics. This limits the total amount of achievable information and results in only a few number of counts of interest area per pixel which makes it crucial to properly evaluate the results from planar surfaces.

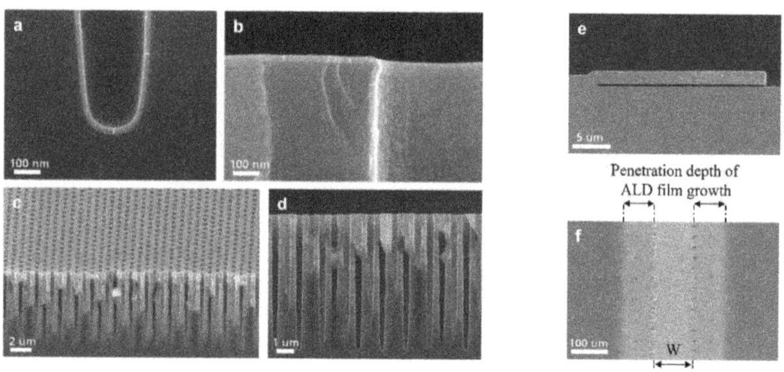

Figure 7. SEM cross-section of 20 nm Si-doped HfO_2 (HSO) deposited on vertical high aspect ratio (VHAR) structures are shown in (**a**–**d**). In (**e**), a side-view SEM micrograph of PillarHall lateral high aspect ratio (LHAR) test structure (membrane width 20 µm) is shown and in (**f**), a top-view SEM micrograph of Si-doped HfO2 grown at 280 °C in LHAR structure (PillarHall) is shown. The membrane has been peeled off (membrane width > 500 µm). The central cavity opening for precursors flow shows with "W" (see Figure 2a). The diffusion depth of the ALD process can be examined directly by ToF-SIMS.

To improve image analysis, we facilitate the characterization procedure of HAR structures on the LHAR to create a suitable technique on imaging elemental distribution in non-planar structures with ToF-SIMS. Figure 7e,f illustrate an LHAR structure via SEM micrographs. The chips (PillarHall) can easily be brought on to a carrier wafer in the ALD chamber. After the deposition process, the membrane is peeled off using an adhesive tape.

Nanomaterials **2019**, *9*, 1035
Nanomaterials **2019**, *9*, x FOR PEER REVIEW 9 of 14

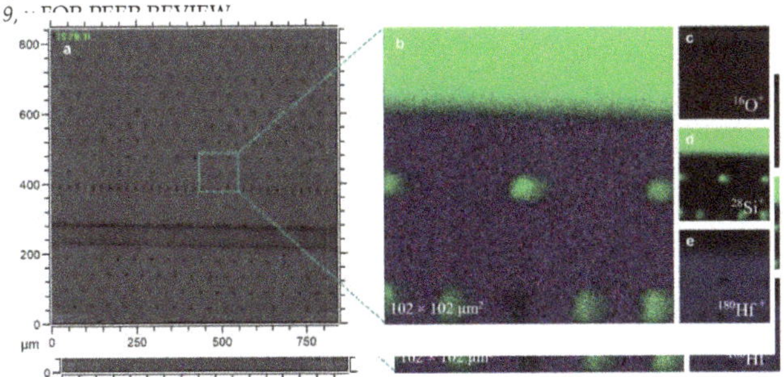

Figure 8. Images showing (**a**) snapshot using a microscope from the PillarHall after membrane removal (top view), (**b**) a superimposed image with ToF-SIMS of deposited region, and (**c**–**e**) an RGB representation of $^{16}O^+$ (m/z = 15.994), $^{28}Si^+$ (m/z = 27.978), and $^{180}Hf^+$ (m/z = 179.949).

Figure 8b represents the total ion image spectrum of the acquired area associated with the $^{16}O^+$, $^{28}Si^+$, and $^{180}Hf^+$ signals. In Figures 8c–e, each elemental signal is color coded in red, green and blue for $^{16}O^+$, $^{28}Si^+$, and $^{180}Hf^+$, respectively. The green dots in Figures 8b and 8d are the holes left by the pillars after membrane removal. Figure 9a illustrates the volumetric view by 3D data analysis of the distribution of $^{16}O^+$ (m/z = 15.994), $^{28}Si^+$ (m/z = 27.978) and $^{180}Hf^+$ (m/z = 179.949) over a 102 × 102 µm² area. To assess the deposited material in the LHAR structure with respect to the depth, the region of interests (ROIs) with a dimension of 2 × 30 µm² were defined as depicted in Figure 9b. For each ROI, the $^{28}Si^+$ and $^{180}Hf^+$ species total counts from emitted secondary ions are integrated, where the influence of the silicon substrate is excluded based on the $^{28}Si^+$ depth profile variation from 16% to 84% on minimum $^{28}Si^+/^{180}Hf^+$ ratio (Figure 10_ROI.01). We are assuming a homogeneous spatial distribution of Si^+ and Hf^+ on each ROI (2 × 30 µm²).

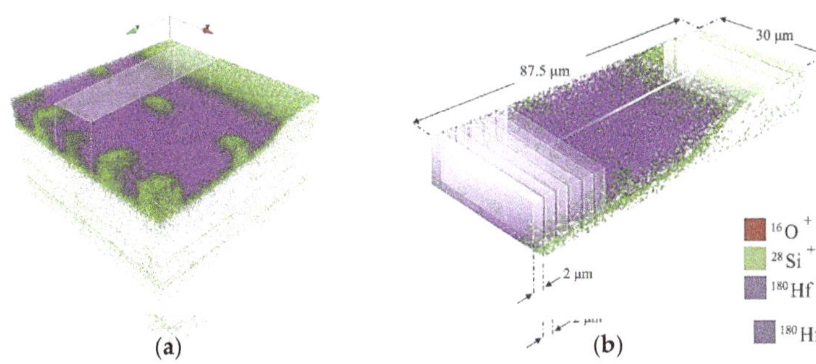

(**a**) (**b**)

Figure 9. (**a**) A volumetric view of the distribution of $^{16}O^+$ (m/z = 15.994), $^{28}Si^+$ (m/z = 27.978) and $^{180}Hf^+$ (m/z = 179.949) over a 102 × 102 µm² field of view from the deposited region after peeling off the membrane. (**b**) A defined ROI with dimension of 2 × 30 µm² to integrate total counts of the interested area from Si and Hf.

The ToF-SIMS analysis results of the HSO material in the LHAR structure are shown in Figure 10. The integrated counts for the $^{180}Hf^+$ and $^{28}Si^+$ species are plotted with respect to the depth inside the test structure. Due to the geometry of the sample, the analysis starts at a minimal distance of 5 µm from the inlet slit. The absolute intensity is not constant as the sputter rate exhibits local variations (not shown here). However, the ratio of the $^{180}Hf^+$ and $^{28}Si^+$ signals is consistent over a depth range from 5 µm to 30 µm.

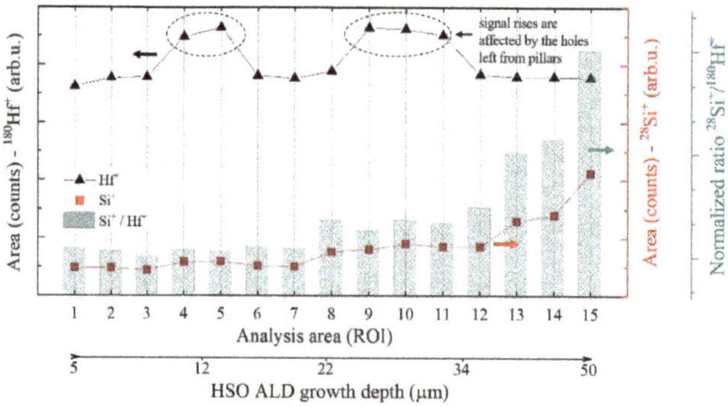

Figure 10. The $^{28}Si^+$, $^{180}Hf^+$ and the ratio signals of the HfO_2 layer obtained by ToF-SIMS analysis of the LHAR test structure with respect to the diffusion depth.

Total area counts of the $^{28}Si^+$ and $^{180}Hf^+$ signals were calculated from the defined areas (ROIs) of the 3D reconstruction (as seen in Figure 9b). A constant $^{28}Si^+$ to $^{180}Hf^+$ signal ratio is observed over the depth range until ~30 µm (deviation of 5%) which corresponds to an aspect ratio up to 1:50 with uniform dopant concentration. At a depth higher than 35 µm, the Si content rises. This can be due to the Si substrate influence. However, up to 30 µm, the step coverage (or depth penetration) of 3DMAS/O_3 (SiO_2) is as good as TEMAHf/O_3 (HfO_2) at the chosen process conditions.

4. Conclusions

This study aimed to illustrate a novel method to address the challenges of elemental analysis in high aspect ratio structures and to provide a practical guideline on the effect of surface topography on ToF-SIMS. As shown, the main difficulty in the interpretation of non-planar structures is to reach optimal material removal and sensitivity for a given material. As shown in the first study, on the vertical TSVs, the image intensity is affected by the surface topography and results in an inaccurate representation of 3D imaging. Since the sputter yield is a function of the primary ion beam incident angle (incident at $\theta = 45°$ to the sample normal for a flat sample), the geometrical shadowing effects are inevitable. In addition, we used lateral high aspect ratio (LHAR) test structures (PillarHall™) as a new platform to perform precise measurements of hafnia doped thin films. In contrast to the common vertical HAR structures, the combination of ToF-SIMS 3D imaging and LHAR test chips leads to a constructed footprint of the composition of the deposited thin film without being influenced by the sputter yield and the surface topography. By integrating the total data points of the ROIs, semi-quantitative calculations on the ratio of elemental distribution over the depth of the LHAR structures were calculated. In conclusion, surface topographic effects could be eliminated by using LHAR structures, allowing the separation of chemical variations within the analysis area. Further investigations are planned to precisely compare the stoichiometry of the deposited material in the planar structures with the deposited film in the depth of the LHAR structures by coupling results from ToF-SIMS and X-ray photoelectron spectroscopy (XPS).

Author Contributions: Investigation and Methodology conducted by A.M.K., C.M. and N.H. Writing the original draft by A.M.K. All authors contributed to the review and editing. LHAR test structures from M.U. and R.L.P. TSVs are provided by S.E. Thin films depositions conducted by S.E. and C.M. ToF-SIMS, SEM and TEM conducted by A.M.K. Results visualization done by A.M.K. The supervision of the project was by N.H. and W.W.

Funding: This research was funded by the German federal ministry of education and research (BMBF) within the project name "VEProSi", No. 13XP5020C and European Regional Development Fund (EFRE) and the Free State of Saxony under the project name "CONSIVA", No. 100273858.

Acknowledgements: The author gratefully thanks Jennifer Emara for reviewing and commenting on the manuscript.

Conflicts of Interest: The authors declare no conflict of interest.

Conflicts of Interest: The authors declare no conflict of interest.

Appendix A

The main challenge of TSV characterization is to gain information from inside of the hole particularly the bottom part. As the TSV dimensions are almost larger than the cutting methods capability, such as FIB cross sectioning, it is hard to design a metrology in order to investigate the depth compositional profile, layer thicknesses, surface quality, conformality and overall step coverage all the way through a hole. The FIB could perform on TSV, but there are some challenges involved, such as sidewall Si redeposition, time consumption in the range of hours as well as low accuracy of a wavy cut known as curtaining effect. Transmission electron microscopy (TEM) with high resolution images gives the most information about deposited layers in a few nanometers range, but it takes lot of time and effort, not only for sample imaging, but also for sample preparation.

According to the aforementioned challenges, for observing the TSV bottom sidewalls with TEM, the sample was cleaved precisely. Therefore, a series of indentations before and after the target structure were applied by a Nano Indenter tool in a way to hit the TSV. Subsequently, the cleaved sample was transferred to the FIB/SEM Strata 400, FEI tool. Figures below show the lamella preparation and lift-out procedure systematically using FIB technique. Carbon and e-Pt protection layers were applied respectively before utilizing any ion beam so as to protect the thin film on the structure (Figure A1a).

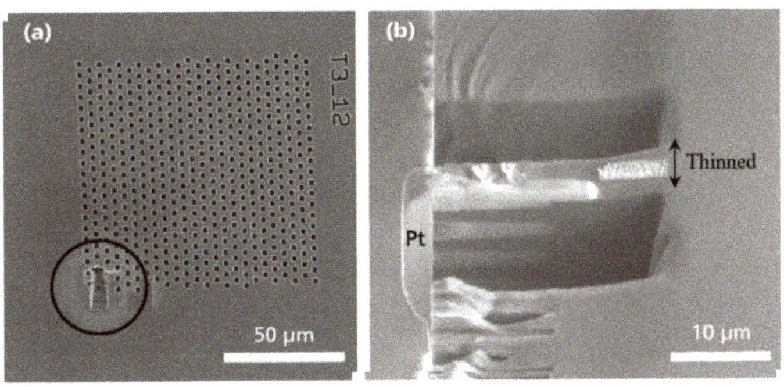

Figure A1. SEM images of (**a**) structure layout and (**b**) SEM images of applied carbon and e-Pt protection layers to the cleaved TSV and rough thinning lamella before lift-out and after tilting.

After depositing carbon, e-Pt and also i-Pt protection layers, the target layer is safe, and it is time to dig in the lamella surroundings. Then, as the lamella is still attached and stable enough, rough thinning has to be done line by line via "cleaning cross section thinning" mode before lift-out (Figure A1b). After the lamella release-cutting at the very bottom, next step is lifting out the free lamella which was welded earlier with Pt to the Omniprobe manipulator (Figure A2a) by GIS.

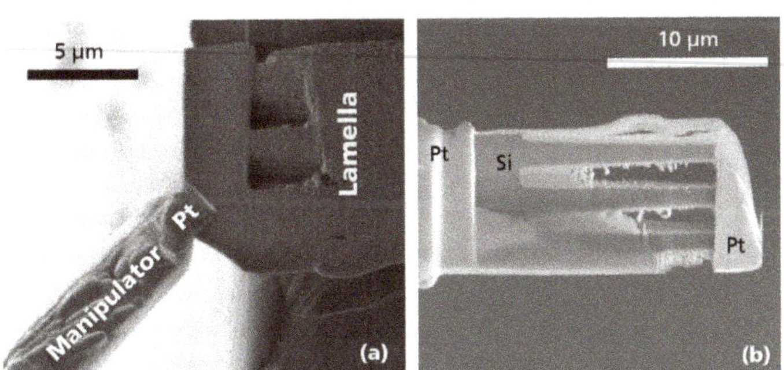

Figure A2. SEM images of (**a**) lifting out the free lamella and (**b**) oblique view during finishing and post-finishing processes with i-Beam and e-Beam.

Figure A2. SEM images of (**a**) lifting out the free lamella and (**b**) oblique view during finishing and post-finishing processes with i-Beam and e-Beam.

Then, the lamella was welded to the TEM sample holder. The manipulator welded area was unattached and finally it was withdrawn. In this stage, finishing and post-finishing processes of the lamella with i-Beam and e-Beam were performed respectively which is shown in Figure A2b. The sample is now thin and transparent enough for TEM observation. Therefore, the welded sample to the sample holder was transferred to the TEM Tecnai F20 200 kV, FEI tool for high resolution pictures (Figure A3).

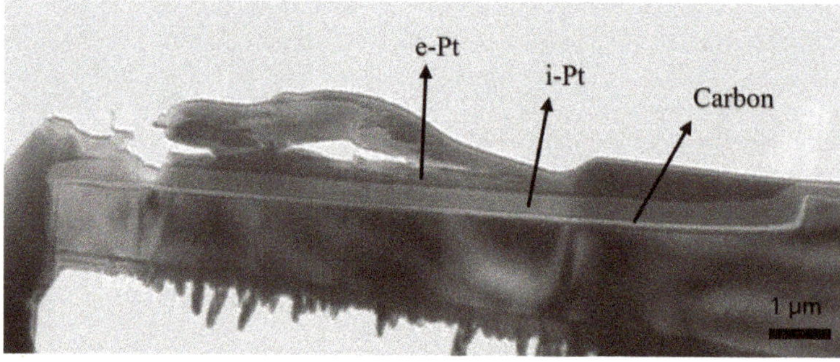

Figure A3. TEM images of the entire overview of halved TSV structure.

References

1. Riviere, J.C.; Myhra, S. *Handbook of Surface and Interface Analysis*, 2nd ed.; CRC Press Taylor & Francis Group: Boca Raton, FL, USA, 2017; pp. 65–96. ISBN 978-1-138-11363-3.
2. Benninghoven, A.; Rüdenauer, F.G.; Werner, H.W. *Secondary Ion Mass Spectrometry: Basic Concepts, Instrumental Aspects, Applications, and Trends*, 1st ed.; J. Wiley: New York, NY, USA, 1987; pp. 435–435. ISBN 0-471-01056-1.
3. Hagenhoff, B. High Resolution Surface Analysis by TOF-SIMS. *Mikrochim. Acta* **2000**, *132*, 259–271.
4. Tortora, L.; Urbini, M.; Fabbri, A.; Branchini, P.; Mariucci, L.; Rapisarda, M.; Barra, M.; Chiarella, F.; Cassinese, A.; Di Capua, F.; et al. Three-dimensional characterization of OTFT on modified hydrophobic flexible polymeric substrate by low energy Cs+ ion sputtering. *Appl. Surf. Sci.* **2018**, *448*, 628–635. [CrossRef]
5. Sameshima, J.; Sugahara, T.; Ishina, T.; Nagao, S.; Suganuma, K. 3D imaging of backside metallization of SiC-SBD influenced by annealing. *J. Mater. Sci. Mater. Electron.* **2019**, *30*, 10848–10856. [CrossRef]
6. Siljeström, S. Single Fluid Inclusion Analysis Using ToF-SIMS Implications for Ancient Earth Biodiversity and Paleoenvironment Studies. Ph.D. Thesis, Stockholm University, Stockholm, Sweden, 2011.
7. Graham, D.J.; Gamble, L.J. Dealing with image shifting in 3D ToF-SIMS depth profiles. *Biointerphases* **2018**, *13*, 06E402. [CrossRef] [PubMed]
8. Oh, H.; Gu, J.M.; Hong, S.J.; May, G.S.; Bakir, M.S. High-aspect ratio through-silicon vias for the integration of microfluidic cooling with 3D microsystems. *Microelectron. Eng.* **2015**, *142*, 30–35. [CrossRef]
9. Taklo, M.M.V.; Klumpp, A.; Ramm, P.; Kwakman, L.; Franz, G.; Taklo, M.M.V. Bonding and TSV in 3D IC Integration: Physical Analysis with a Plasma FIB. *Microsc. Microanal.* **2011**, *25*, 9–12.
10. Armini, S.; El-Mekki, Z.; Vandersmissen, K.; Philipsen, H.; Rodet, S.; Honore, M.; Radisic, A.; Civale, Y.; Beyne, E.; Leunissen, L. Void-Free Filling of HAR TSVs Using a Wet Alkaline Cu Seed on CVD Co as a Replacement for PVD Cu Seed. *J. Electrochem. Soc.* **2010**, *158*, H160–H165. [CrossRef]
11. Esmaeili, S.; Lilienthal, K.; Nagy, N.; Gerlich, L.; Krause, R.; Uhlig, B. Co-MOCVD processed seed layer for through silicon via copper metallization. *Microelectron. Eng.* **2019**, *211*, 55–59. [CrossRef]
12. Müller, J.; Yurchuk, E.; Schlösser, T.; Paul, J.; Hoffmann, R.; Müller, S.; Martin, D.; Slesazeck, S.; Polakowski, P.; Sundqvist, J.; et al. Ferroelectricity in HfO2 enables nonvolatile data storage in 28 nm HKMG. In Proceedings of the 2012 Symposium on VLSI Technology (VLSIT), Honolulu, HI, USA, 12–14 June 2012; pp. 25–26. [CrossRef]
13. Ali, T.; Polakowski, P.; Riedel, S.; Büttner, T.; Kämpfe, T.; Rudolph, M.; Pätzold, B.; Seidel, K.; Löhr, D.; Hoffmann, R.; et al. High Endurance Ferroelectric Hafnium Oxide-Based FeFET Memory Without Retention Penalty. *IEEE Trans. Electron Devices* **2018**, *65*, 3769–3774. [CrossRef]

14. Krivokapic, Z.; Rana, U.; Galatage, R.; Razavieh, A.; Aziz, A.; Liu, J.; Shi, J.; Kim, H.J.; Sporer, R.; Serrao, C.; et al. 14nm ferroelectric FinFET technology with steep subthreshold slope for ultra low power applications. In Proceedings of the 2017 IEEE International Electron Devices Meeting (IEDM), San Francisco, CA, USA, 2–6 December 2017; pp. 15.1.1–15.1.4. [CrossRef]
15. Kobayashi, M.; Hiramoto, T. On device design for steep-slope negative-capacitance field-effect-transistor operating at sub-0.2V supply voltage with ferroelectric HfO_2 thin film. *AIP Adv.* **2016**, *6*, 025113. [CrossRef]
16. Kirbach, S.; Kühnel, K.; Weinreich, W. Piezoelectric Hafnium Oxide Thin Films for Energy-Harvesting Applications. In Proceedings of the 2018 IEEE 18th International Conference on Nanotechnology (IEEE-NANO), Cork, Ireland, 23–26 July 2018; pp. 31–34. [CrossRef]
17. Kühnel, K.; Czernohorsky, M.; Mart, C.; Weinreich, W. High-density energy storage in Si-doped hafnium oxide thin films on area-enhanced substrates. *J. Vac. Sci. Technol. B* **2019**, *37*, 021401. [CrossRef]
18. Mart, C.; Kämpfe, T.; Zybell, S.; Weinreich, W. Layer thickness scaling and wake-up effect of pyroelectric response in Si-doped HfO_2. *Appl. Phys. Lett.* **2018**, *112*, 052905. [CrossRef]
19. Anderle, M.; Barozzi, M.; Bersani, M.; Giubertoni, D.; Lazzeri, P. Ultra Shallow Depth Profiling by Secondary Ion Mass Spectrometry Techniques. *AIP Conf. Proc.* **2003**, *695*, 695–704. [CrossRef]
20. Acartürk, T.; Semmelroth, K.; Pensl, G.; Saddow, S.E.; Starke, U. Concentration of N and P in SiC Investigated by Time-Of-Flight Secondary Ion Mass Spectrometry (TOF-SIMS). *Mater. Sci. Forum* **2009**, *483–485*, 453–456. [CrossRef]
21. Gao, F.; Arpiainen, S.; Puurunen, R.L. Microscopic silicon-based lateral high-aspect-ratio structures for thin film conformality analysis. *J. Vac. Sci. Technol. A* **2014**, *33*, 010601–010605. [CrossRef]
22. Ylilammi, M.; Ylivaara, O.M.E.; Puurunen, R.L. Modeling growth kinetics of thin films made by atomic layer deposition in lateral high-aspect-ratio structures. *J. Appl. Phys.* **2018**, *123*, 205301. [CrossRef]
23. Cremers, V.; Puurunen, R.L.; Dendooven, J. Conformality in atomic layer deposition: Current status overview of analysis and modelling. *Appl. Phys. Rev.* **2019**, *6*, 21302. [CrossRef]
24. Puurunen, R.L. Surface chemistry of atomic layer deposition: A case study for the trimethylaluminum / water process. *J. Appl. Phys.* **2005**, *97*, 121301. [CrossRef]
25. Ahern, P. Time-of-Flight Secondary Ion Mass Spectrometry (ToF-SIMS)—Theory and Practice. 2013. Available online: http://www.linkedin.com/in/paulahern1 (accessed on 29 April 2019).
26. Kozole, J.; Willingham, D.; Winograd, N. The effect of incident angle on the $C60^+$ bombardment of molecular solids. *Appl. Surf. Sci.* **2008**, *255*, 1068–1070. [CrossRef] [PubMed]
27. Green, F.M.; Gilmore, I.S.; Seah, M.P. TOF-SIMS: Accurate mass scale calibration. *J. Am. Soc. Mass Spectrom.* **2006**, *17*, 514–523. [CrossRef] [PubMed]
28. Lee, J. Time-of-Flight Secondary Ion Mass Spectrometry—Fundamental Issues for Quantitative Measurements and Multivariate Data Analysis. Ph.D. Thesis, Oxford University, Oxford, UK, 2011.
29. Lee, J.L.S.; Gilmore, I.S.; Seah, M.P.; Fletcher, I.W. Topography and Field Effects in Secondary Ion Mass Spectrometry—Part I: Conducting Samples. *J. Am. Soc. Mass Spectrom.* **2011**, *22*, 1718–1728. [CrossRef] [PubMed]
30. Polakowski, P.; Riedel, S.; Weinreich, W.; Rudolf, M.; Sundqvist, J.; Seidel, K.; Muller, J. Ferroelectric deep trench capacitors based on Al:HfO_2 for 3D nonvolatile memory applications. In Proceedings of the 2014 IEEE 6th International Memory Workshop (IMW), Taipei, Taiwan, 18–21 May 2014. [CrossRef]
31. Mart, C.; Weinreich, W.; Czernohorsky, M.; Riedel, S.; Zybell, S.; Kuhnel, K. CMOS Compatible pyroelectric applications enabled by doped HfO_2 films on deep-trench structures. In Proceedings of the 48th European Solid-State Device Research Conference (ESSDERC), Dresden, Germany, 3–6 September 2018; pp. 130–133. [CrossRef]
32. Richter, C.; Schenk, T.; Park, M.H.; Tscharntke, F.A.; Grimley, E.D.; LeBeau, J.M.; Zhou, C.; Fancher, C.M.; Jones, J.L.; Mikolajick, T.; et al. Si Doped Hafnium Oxide—A 'Fragile' Ferroelectric System. *Adv. Electron. Mater.* **2017**, *3*, 1700131. [CrossRef]
33. Ali, T.; Polakowski, P.; Riedel, S.; Büttner, T.; Kämpfe, T.; Rudolph, M.; Pätzold, B.; Seidel, K.; Löhr, D.; Hoffmann, R.; et al. Silicon doped hafnium oxide (HSO) and hafnium zirconium oxide (HZO) based FeFET: A material relation to device physics. *Appl. Phys. Lett.* **2018**, *112*, 222903. [CrossRef]
34. Piehowski, P.D.; Davey, A.M.; Kurczy, M.E.; Sheets, E.D.; Winograd, N.; Ewing, A.G.; Heien, M.L. Time-of-Flight Secondary Ion Mass Spectrometry Imaging of Subcellular Lipid Heterogeneity: Poisson Counting and Spatial Resolution. *Anal. Chem.* **2009**, *81*, 5593–5602. [CrossRef] [PubMed]

35. Vanbellingen, Q.P.; Elie, N.; Eller, M.J.; Della-Negra, S.; Touboul, D.; Brunelle, A. Time-of-flight secondary ion mass spectrometry imaging of biological samples with delayed extraction for high mass and high spatial resolutions. *Rapid Commun. Mass Spectrom.* **2015**, *29*, 1187–1195. [CrossRef] [PubMed]
36. Sodhi, R.N.S. Time-of-flight secondary ion mass spectrometry (TOF-SIMS): - Versatility in chemical and imaging surface analysis. *Analyst* **2004**, *129*, 483–487. [CrossRef] [PubMed]

nanomaterials

MDPI

Article

Role of Structure and Composition on the Performances of P-Type Tin Oxide Thin-Film Transistors Processed at Low-Temperatures

Raquel Barros [1,2], Kachirayil J. Saji [1,3], João C. Waerenborgh [4], Pedro Barquinha [1], Luís Pereira [1], Emanuel Carlos [1], Rodrigo Martins [1,*] and Elvira Fortunato [1,*]

1 CENIMAT/I3N, Departamento de Ciência dos Materiais, Faculdade de Ciências e Tecnologia, FCT, Universidade Nova de Lisboa and CEMOP-UNINOVA, Campus da Caparica, 2829-516 Caparica, Portugal; rbarros@hovione.com (R.B.); saji@cusat.ac.in (K.J.S.); pmcb@fct.unl.pt (P.B.); lmnp@fct.unl.pt (L.P.); e.carlos@campus.fct.unl.pt (E.C.)

2 Hovione, Campus do Lumiar, Edifício S, Estrada do Paço do Lumiar, 1649-038 Lisboa, Portugal

3 International School of Photonics, Cochin University of Science and Technology, Kochi – 682 022, India

4 C2TN, DECN, Instituto Superior Técnico, Universidade de Lisboa, 2695-066 Bobadela LRS, Portugal; jcarlos@ctn.tecnico.ulisboa.pt

* Correspondence: rm@uninova.pt (R.M.); emf@fct.unl.pt (E.F.)

Received: 30 December 2018; Accepted: 21 February 2019; Published: 1 March 2019

Abstract: This work reports on the role of structure and composition on the determination of the performances of p-type SnO_x TFTs with a bottom gate configuration deposited by rf magnetron sputtering at room temperature, followed by a post-annealed step up to 200 °C at different oxygen partial pressures (O_{pp}) between 0% and 20% but where the p-type conduction was only observed between in a narrow window, from 2.8% to 3.8%. The role of structure and composition were evaluated by XRD and Mössbauer spectroscopic studies that allows to identify the best phases/compositions and thicknesses (around 12 nm) to be used to produce p-type TFTs with saturation mobility of 4.6 $cm^2\ V^{-1}\ s^{-1}$ and on-off ratio above 7×10^4, operating at the enhancement mode with a saturation voltage of −10 V. Moreover, a brief overview is also presented concerning the present state of the existing developments in processing SnO_x TFTs with different methods and using different device configurations.

Keywords: p-type TFT; p-type oxide semiconductors; SnO electrical properties; oxide structure analysis

1. Introduction

Oxide electronics are a promising alternative to amorphous silicon (a-Si:H) and organic semiconductors to build reliable Thin Film Transistors (TFT) and more complex electronic circuits, addressing the challenges of flexible electronics and of the low cost and disposable electronics. In spite of the earlier work made during the 60s concerning the processing of n-type TFT [1,2], only forty five years later, with the work of Hosono [3], Wager [4], Carcia [5] and Fortunato [6], a significant worldwide interest materialised, especially for the active matrix for organic light emitting diodes (AMOLED) technology, exploiting their electronic properties, such as high saturation mobility, excellent uniformity and homogeneity, together with a high reliability associated with a low or room processing temperature [7].

However, there is no report on p-type oxide TFTs that are processed and cured at low temperatures with a performance similar to n- type, due to the low hole mobilities so far achieved in running stable and reliable devices [8]. The achievement of reliable p-type TFT, with performances similar to n-type TFT is of great importance for shaping electronics challenges towards the production of complementary metal oxide semiconductors (CMOS), a key device for analogic and digital electronic systems, thanks to their low power consumption. This is a noticeable relevant CMOS property for low cost flexible electronics. To this end, we could use organic semiconductors, aiming to exploit the advantage that they can be processed at low temperatures.

Concerning organic p-type TFTs, most device performances on stability and mobility (<2 cm^2 V^{-1} s^{-1}) are low [9–12], while the n-type organic TFT still exhibits low mobilities ($\leq$1 cm^2 V^{-1} s^{-1}) and requires a high absolute on voltages to switch it on [13,14].

An alternative to this is the inorganic oxide TFT, which is robust but in most cases requires high process temperatures. So far, most of the reported oxide TFTs are n-type, processed either on rigid or flexible substrates in which exists a consolidate set of results for films processed via physical or chemical methods [1–8,15,16]. For p-type, the transport due to holes is associated wirh oxygen *p* asymmetric orbitals, which severely limit the carrier mobility and therefore the TFT performances. In spite of Cu_2O being a p-type oxide with mobility >100 cm^2 V^{-1} s^{-1} [17,18], the TFT based on these thin films or their compositions as Cu:NiO, exhibit mobilities and On-Off current ratios of <1.5 cm^2 V^{-1} s^{-1} and 10^4 respectively [17–21]. Other materials have been also reported, such as NiO_x processed/annealed at 300 °C, exhibiting mobilities above 25 cm^2 V^{-1} s^{-1} [22].

Tin oxide has been studied as an alternative material to produce p-type oxides, with similar performances as those obtained in n-type oxides. The structure, morphology and ambipolar characteristic of these films are well known for oxides processed by reactive sputtering using metal targets and heat treated at 400 °C [23]. Indeed, it is known that SnO has an indirect band gap structure specifically controlled by the divalent tin (SnII), in a layered crystal structure [24,25] with major contributions from Sn 5*s* and O 2*p* orbitals near the valence band maximum (VBM) and Sn 5*p* orbitals towards the conduction band minimum (CBM). The p-type behaviour is mainly attributed to the Sn vacancy and the O interstitial where tin is in Sn^{2+} oxidation state [24,25]. The excess oxygen in the film transforms some cations in Sn^{3+} to maintain electrical neutrality. This process is considered to be Sn^{2+} capturing a hole and forming weak bonded holes, located inside the bandgap, near the top of the valence band as localized acceptor states [26,27]. This means that the final free carriers' behaviour of the films process is highly dependent on how oxygen is bonded and how it may compensate for defects.

Here, the contributions from Sn 5*s* states to VBM offer appreciable hole mobility in this material, without using a high processes temperature [28,29]. This leads also to the production of TFT with different geometry configurations [30] or using, besides metallic targets, ceramic ones on films grown by rf magnetron sputtering, heat treated at 400 °C [31].

In Table 1 we present the set of developments obtained concerning the performances of p-type SnO TFTs produced by Radio Frequency Magnetron Sputtering (RFMS) in the last 10 years [28,32–49]. There, we also present the architecture selected (SBG: staggered bottom-gate; STG: staggered top-gate; CBG: coplanar bottom-gate; CTG: coplanar top-gate; DG: double-gate), the process temperature, the oxygen partial pressure (O_{pp}) and the type of dielectric used.

Table 1. Recent developments concerning the performances of p-type SnO TFTs processed by Radio Frequency Magnetron Sputtering (RFMS), using different type of device configurations, O_{pp}, dielectrics and process temperatures.

Method	Device Structure	Process Temp. (°C)	O_{PP} (%)	Dielectric	$\mu_{h,FE}$ ($cm^2\ V^{-1}\ s^{-1}$)	I_{on}/I_{off}	Year	Ref.
RFMS	SBG	RT *	11.5	ATO	1.2	10^3	2010	[28]
RFMS	SBG	300	1	SiN_x	0.24	10^2	2010	[32]
RFMS	SBG	150	n.r.	Paper	1.3	10^2	2011	[33]
RFMS	SBG	250	n.r.	SiO_2	1.8	10^3	2013	[34]
RFMS	SBG	225	4.3	HfO_2	0.33	10^3	2014	[35]
RFMS	SBG	230	n.r.	SiO_2	0.59	3×10^3	2014	[36]
RFMS	CTG	200	9	P(VDF-TrFE)	3.3	3×10^2	2014	[37]
RFMS	SBG	200	11.8	SiO_2	1.36	2×10^3	2014	[38]
RFMS	STG	200	9	P(VDF-TrFE)	2.7	2×10^2	2014	[39]
RFMS	SBG	200	n.r.	SiO_2	0.61	6.2×10^5	2015	[40]
RFMS	SBG	250	n.r.	SiO_2	1.8	10^5	2015	[41]
RFMS	SBG	200	7.5	SiO_2	4.13	6×10^2	2015	[42]
RFMS	SBG	250	-	SiO_2	1.16	2.3×10^2	2015	[43]
RFMS	SBG	300	n.r.	SiO_2	3.33	10^4	2016	[44]
RFMS	SBG	200	n.r.	SiO_2	0.63	5.2×10^6	2017	[45]
RFMS	SBG	225	n.r.	Al_2O_3	0.7	2.6×10^4	2018	[46]
RFMS	SBG	120	9	HfO_2	5.53	2.7×10^3	2018	[47]
RFMS	SBG	225	3.1	SiO_2	1.41	1.5×10^3	2018	[48]
RFMS	SBG	225	3.1	SiO_2	0.87	1.88×10^4	2019	[49]
RFMS	SBG	RT *	3.0	ATO	4.6	7×10^4	2019	This work

* Post deposition annealed at 200 °C; n.r.: not reported; RT = Room Temperature; ATO = Aluminum Titanium Oxide.

Overall, we notice that the only devices processed at room temperature using the RFMS technique are those developed by the present group [28,33]. Here, it is also relevant to mention that the presence of low oxygen partial pressure during the deposition process enables the production of more stable devices [35,48,49]. Apart from that, the configuration most used is the staggered bottom-gate, while the most common dielectric used is the silicon dioxide. Apart from that, most of the substrates used are rigid (glass or silicon wafer), except that referred to as CMOS devices integrating p-type TFT based on SnO_x made on paper [33].

Besides stability and reproducibility issues, the data presented show that the device with the best mobility (5.53 $cm^2\ V^{-1}\ s^{-1}$, with Perovskite-Mediated Photogating [47]) does not correspond to the device with the highest On/Off (I_{on}/I_{off}) ratio (5.2×10^6, using argon-plasma surface treatment [45]). Apart from that, most of the TFT studied does not work on the enhancement mode, as desired for application purposes.

Moreover, we noticed that the thickness of the channel layer, together with the state of the surface (degree of roughness and surface defects), determine the electrical characteristics presented by TFT and its stability.

In Table 2 we present the most significant data achieved in the last ten years concerning the production of SnO_x p-type TFT using different processing techniques such as: Pulsed Laser Deposition (PLD); Electron-Beam Evaporation (EBE); Thermal Evaporation (TE); direct current magnetron sputtering (DCMS); PVD: Physical Vapor Deposition (PVD); Atomic Layer Deposition (ALD); Spin-Coating (SC). As in Table 1, the different type of device configurations are also shown (SBG: staggered bottom-gate; STG: staggered top-gate; CBG: coplanar bottom-gate; CTG: coplanar top-gate; DG: double-gate); Oxygen partial pressures (O_{pp}); dielectrics and process temperatures used.

Table 2. Recent developments concerning the performances of p-type SnO TFTs processed by other physical and chemical process techniques, for different device configurations, O_{pp}, dielectrics and process temperatures.

Method	Device Structure	Process Temp. (°C)	Oxygen Partial Pressure (%)	Dielectric	$\mu_{h,FE}$ ($cm^2\ V^{-1}\ s^{-1}$)	I_{on}/I_{off}	Year	Ref.
PLD	STG	575	4×10^{-2} Pa	Al_2O_3	1.3	10^2	2008	[50]
PLD	STG	250	1×10^{-2} Pa	SiO_2	0.81	~10^2	2011	[51]
PLD	SBG	300	1×10^{-2} Pa	SiO_2	2.18	–	2014	[52]
EBE	SBG	400	n.r.	SiO_2	0.32	5×10^2	2013	[53]
TE	SBG	250	n.r.	Al_2O_3	1.4	5×10^4	2018	[54]
DCMS	SBG	180	9	HfO_2	6.75	~10^3	2013	[29]
DCMS	DG	300	3.07×10^{-2} Pa	SiO_2	6.54	10^5	2015	[30]
DCMS	SBG	200	n.r.	HfO_2	5.56	4.8×10^4	2016	[55]
DCMS	SBG	200	n.r.	HfO_2	7.6	3×10^4	2018	[31]
PVD	SBG	200	n.r.	HfO_2	2.13	9.6×10^6	2017	[56]
ALD	SBG	250	n.r.	Al_2O_3	1	2×10^6	2017	[57]
SC	SBG	450	-	SiO_2	0.13	85	2012	[58]

* Post deposition annealed at 200 °C; n.r.: not reported.

Overall, the best p-type TFTs fabricated so far have been those processed by DCMS, exhibiting a mobility of 6.54 $cm^2\ V^{-1}\ s^{-1}$ and an On/Off ratio of 10^5, working in the depletion mode [30]. Moreover, the p-type TFT processed by PVD and using a STG configuration exhibit the highest recorded On/Off ratio (9.6×10^6) [56].

From the present state of the art, we saw that there are several parameters that impact on the electrical performance presented by p-type TFT SnO_x based, most of them connected to the process parameters used, the structure of the films obtained, as well as the dielectric and the geometry configuration used.

In this paper, we report the fabrication of p-type SnO_x TFTs deposited by RFMS technique at RT that are post-annealed up to 200 °C, turning the process compatible with the use of low-cost flexible substrates as paper [7]. In this study, we aim to better understand the role that the structure, surface finishing and oxygen play during the growing process of SnO_x in order to define a process window that allows the production of reliable and high stable p-type TFT with high electronic performances, such as field effect mobility and On-Off- current ratios.

2. Materials, Methods and Results

2.1. Experimental Details

Films Preparation

SnO_x thin films (5–100 nm) were deposited on glass substrates with an r.f. magnetron sputtering system at room temperature, using a metallic tin target (99.999% pure). Depositions were carried out in a controlled atmosphere of oxygen and argon, using an r.f. power of 40 W and 4 substrates of 1 inch ×1 inch placed in the substrate holder were rotated at a speed of 40 rpm, aiming to get high uniform films overall substrate area. Experiments were performed by varying O_{pp} ($O_{pp} = P_{O2}/(P_{O2} + P_{Ar})$, between 0% and 20%, where P_{O2} and P_{Ar} are partial pressures of oxygen and argon, respectively, keeping the total deposition pressure constant at 0.2 Pa. Moreover, the argon gas flow was kept constant—around 50sccm—while the oxygen gas flow varied from 0 to 12.5 sccm. At these conditions, the deposition rate was 40 Å/min. After deposition, the films were annealed under standard environment conditions at temperatures around 200 °C, for different times inside a tubular furnace.

2.2. Structure Morphology, Composition and Electro-Optical Data and Analysis

Prior to processing the TFT devices, the material in which the channel is based was deposited on glass substrates and their structure, morphology, composition and electro-optical properties were fully analysed, aiming to determine the best conditions in which to grow the channel layers of the TFT. The structure of the films was studied by X-ray diffraction (XRD) using a PANalytical X'Pert PRO (Cambridge, MA, USA) with Cu Kα radiation (λ = 1.540598 Å) while the morphology was assessed by scanning electron microscopy (SEM) with a ZEISS SEM/FIB AURIGA (Jena, Germany) operated at 2 kV, with an aperture size of 30 μm and a working distance of 5.2 nm. The surface roughness of the films was analysed using atomic force microscopy (AFM) with an Asylum MFP-3D instrument (Oxford Instruments, Oxford, UK) in non-contact mode.

The optical transmittance (T%) was measured between 300 to 2500 nm, using a double-beam UV–vis-NIR spectrometer (Lambda 950, San Dimas, CA, USA).

The electrical resistivity (ρ), Hall mobility (μ) and free carrier concentration and their nature (electrons or holes) were determined by Hall effect measurements in Van der Pauw geometry in a Biorad HL 5500 equipment (York, England) using a constant magnetic field of 0.5 T. The electrical properties of the samples were measured at room temperature.

2.3. Structure Data and Analysis

Figure 1 shows the XRD and Mössbauer and CEMS spectra of as-deposited and annealed SnO_x films with O_{PP} = 3.0% and O_{PP} = 3.6% O_{PP}, respectively, as the limits of the interval where a p-type transport behaviour is observed, as proven by Hall effect measurements (a positive Hall coefficient obtained for all samples evaluated, after annealing).

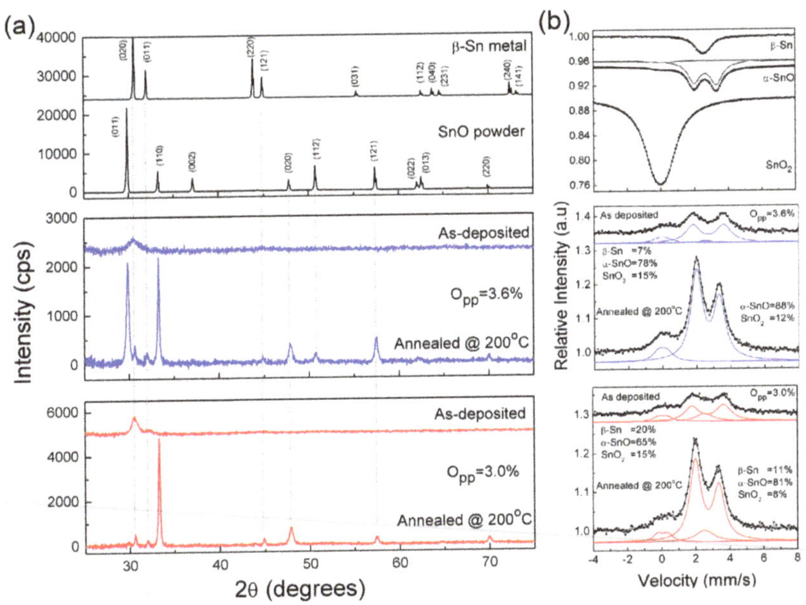

Figure 1. (**a**) Diffractograms of SnO_x thin films as-deposited. (**b**) Transmission ^{119}Sn Mössbauer and CEMS spectra of as-deposited and annealed SnO_x films with O_{PP} = 3.0% and O_{PP} = 3.6% O_{PP}, respectively.

The XRD data show that the films as deposited are amorphous, turning crystalline after annealing at 200 °C.

Figure 1a shows that the XRD pattern of films are 120 nm thick as deposited, where we can see that the metallic tin dominates over the SnO phase. It also shows the XRD diffractograms for SnO powder and metallic Sn, to be taken as references.

Room temperature transmission Mössbauer and ^{119}Sn conversion-electron Mössbauer spectroscopy (CEMS) were performed on the set of samples prepared, before and after annealing, using a proportional backscatter detector RIKON-5 (Wissel) in flowing 5% CH_4-95% He gas mixture (see Figure 1b). The spectra were collected using a conventional constant acceleration spectrometer and a 5 mCi $Ca^{119m}SnO_3$ source. The velocity scale was calibrated using a ^{57}Co (Rh) source and an α-Fe foil. The Sn isomer shifts (IS) are given relative to $BaSnO_3$ reference material at 295 K (RT) and obtained by adding 0.031 mm/s to the IS relative to the source. The spectra were fitted to Lorentzian lines using a non-linear least-squares method. The set of extrapolated parameters extracted are presented in Table 3.

Table 3. Extrapolated parameters from the transmission ^{119}Sn Mössbauer spectra of bulk samples and CEMS spectra of films taken at RT.

Sample	IS (mm/s)	QS (mm/s)	Γ (mm/s)	Sn Phase	I (%)
SnO_2 bulk	0.01	0.56	1.34	-	100
α-SnO bulk	2.67	1.34	0.98	α-SnO	95
	−0.03	0.58	0.77	SnO_2	5
-Sn metal	2.56	-	1.03	-	100
Film O_{pp} = 3.0%, RT	2.76	1.87	1.04	SnO	65
	0.11	0.52	0.84	SnO_2	15
	2.56	–	1.3	-Sn	20
Film O_{pp} = 3.0%, 200 °C, 30 min	2.7	1.38	0.79	SnO	81
	0.08	0.59	0.7	SnO_2	8
	2.56	–	1.35	-Sn	11
Film O_{pp} = 3.6%, RT	2.73	1.82	1.02	SnO	78
	0.11	0.53	0.81	SnO_2	15
	2.56	–	1.3	-Sn	7
Film O_{pp} = 3.6%, 200 °C, 30 min	2.7	1.34	0.82	SnO	88
	0.06	0.48	0.85	SnO_2	12

IS (mm/s) isomer shift relative to metallic $BaSnO_3$ at 295 K; QS (mm/s) quadrupole splitting; Γ (mm/s) line-width; I relative area. Estimated errors ≤0.02 mm/s for IS, QS, Γ and <2% for I.

The Mössbauer spectra of Sn, SnO and SnO_2 samples were also taken as reference samples in order to compare with those of the phases detected in the films by CEMS. The spectra of the reference samples reveal the typical spectra corresponding to β-Sn, α-SnO and SnO_2, respectively [59,60]. The α-SnO spectrum reveals the typical air contamination due to the higher recoilless fraction of Sn^{4+} in SnO_2 as compared to Sn^{2+} in α-SnO [61,62], besides the presence of the α-SnO [60–63]. The spectra recorded were fitted by three contributions connected to isomer shifts (IS) of Sn^{4+}, Sn^{2+} and metallic Sn (see Table 1). The absorption peak due to metallic Sn is similar to that of the β-Sn, confirming its presence, which agreed with the XRD data. Moreover, the IS quadrupole splitting (QS) and the line widths of Sn^{2+} in the films before annealing are higher than the corresponding parameters for bulk α-SnO. After annealing, the widths of Sn^{2+} decrease, reaching values close to those of crystalline α-SnO. This suggests that the Sn^{2+} oxide present is amorphous before annealing, transforming into the crystalline form after annealing at 200 °C for at least 30 minutes. These data agree with those obtained from XRD for the same samples, showing that the films as-deposited are mainly composed of amorphous SnO and metallic β-Sn, with residual amounts of SnO_2, which were only detected by CEMS. The SnO_2 IS and quadrupole splitting (QS) deduced differs from those of bulk SnO_2 which we attribute to the low degree of crystallinity of the films. Assuming that the recoilless factors of β-Sn, SnO and SnO_2 in films are not different for the same species in the different samples, the fraction of Sn atoms in each phase should follow the same trend, with annealing or with O_{pp} used. Indeed, the data recorded reveal that the fraction of Sn present as β-Sn is lower in the film deposited at higher O_{pp} (3.6%), than in those deposited at 3.0%. After annealing at 200 °C, the films crystallize leading to the formation of a strong α-SnO phase which also contributes to the oxidation of β-Sn. Under these conditions, the SnO_x with $1 < x < 2$, is the dominant phase of the channel layer with a small contribution from metallic tin, explaining the p-type transport behaviour observed.

2.4. Electrical Data and Analysis

Hall Effect measurements were performed to identify the charge carrier and carrier mobility in the material. As we are in the presence of ambipolar material, as it is the case of SnO_x [64], electrons and holes will pile up at the same side of the sample and consequently the measured Hall voltage depends on the relative mobilities and concentrations of holes and electrons. Hall mobility for an ambipolar semiconductor is thus given by,

$$\mu_{Hall} = \frac{n \cdot \mu_n^2 - p \cdot \mu_p^2}{n \cdot \mu_n + p \cdot \mu_p} \tag{1}$$

where n, p, μ_n and μ_p represent electron density, hole density, electron mobility and hole mobility respectively. This leads to a reduction in the Hall mobility, compared to the mobilities of the charge carriers.

In the present study, the samples as deposited exhibit fluctuations in the sign and magnitude of the Hall coefficient, where the average mobilities of carriers were of about 10^{-1} cm^2 V^{-1} s^{-1}. After annealing up to 200 °C for 30 minutes, films prepared with 2.8% < O_{pp} <3.8 show a positive Hall coefficient. This suggests a considerably large density of holes compared to the density of electrons, resulting in a positive Hall voltage with typical Hall mobility of 2 cm^2 V^{-1} s^{-1} associated to the materials' evaluated.

Figure 2 shows the resistivity variation of SnO_x films for different O_{pp}, as deposited and after annealing at 200 °C, for different annealing times.

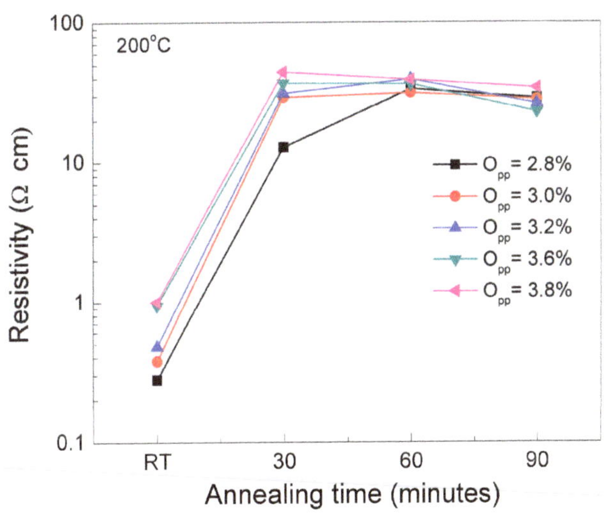

Figure 2. Resistivity of SnO_x films as a function of annealing time, for films processed at different O_{pp}, as depicted in the Figure.

The data show that the resistivity of the films processed tends to saturate around 30–40 Ω·cm, for annealing times above 30 mn. The lowest annealing time is for samples processed with O_{pp} above 3.20%, while the highest time (above 60 mn), for samples prepared with 2.80%< O_{pp} <3.20%.

Annealing at temperatures above 200 °C cause again a decreasing tendency in material resistivity (not shown here) that can be associated with the phase transformation of the material from SnO (p-type) to SnO_2 (n-type). These data are consistent with those depicted in Figure 1b.

2.5. Optical Data and Analysis

Figure 3a shows the optical transmittance data of the films processed for 2.80%< O_{pp} <3.20%, as deposited (RT) and after annealing at 200 °C, during 30 mn. The data depicted show that as O_{pp} increases the films become more transparent, as expected. As deposited, independent of O_{pp} used, the data depicted show, on average, transmittances below 20% in the visible region. These low values are attributed to the presence of large concentration of metallic tin in the films. By annealing the films, the optical transmittance in the visible region increases up to 55%, function of the O_{pp} used.

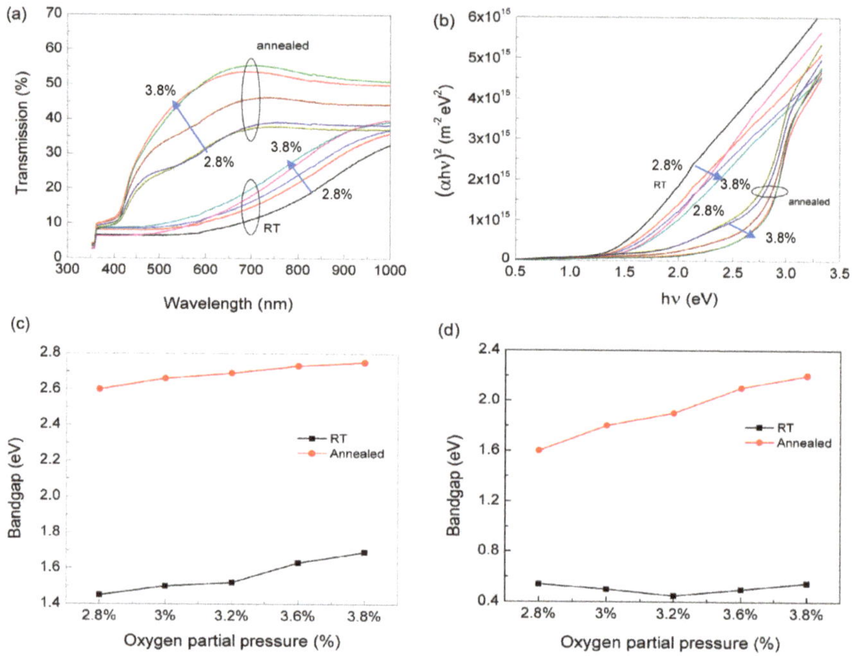

Figure 3. (**a**) Optical transmission of as prepared and annealed SnO_x films. (**b**) Plot for the estimation of the direct band gap estimation; Dependence of the estimated direct (**c**) and indirect (**d**) band gap on O_{pp} used, before and after annealing the films for 30 mn.

The optical band gap E_g of the films were determined by the relationship:

$$\alpha \cdot h \cdot \nu = const \cdot (h \cdot \nu - E_g)^r \quad (2)$$

where α is the absorption coefficient, *hν* denotes the photon energy and *r* a constant depending on the type of optical transition expected. The E_g value is then obtained by linearly extrapolating the plot of $(\alpha \cdot h \cdot \nu)^{1/r}$ versus $h \cdot \nu$ and finding the intersection with the abscissa.

The optical data in Figure 3a were analysed with $r = \frac{1}{2}$ (direct transition, for SnO, expected to be E_g ~2.5 eV and for SnO_2 E_g ~3.6 eV, see Figure 3b) and $r = 2$ (Indirect transition, for SnO E_g ~1 eV). For the different O_{pp} used, as deposited the direct band gap varies between 1.5 eV and 1.8 eV, while the estimated indirect band gap is kept around 0.6 eV. These values reflect the quasi metallic state of the films produced. After annealing, the films are better oxidized and the structure changes, as observed in Figure 1. Overall, we estimate a direct band gap with 2.6 eV <E_g <2.75 eV, while the indirect band gap is 1.6 eV < E_g < 2.2 eV (see Figure 3c,d). These values are close to the reported bandgap values for SnO films, showing for the range of O_{pp} used the SnO_2 phase does not dominate the optical characteristics of the films produced. Small variations in these values could be understood on the basis of fractional variations of various phases (metallic tin, SnO and SnO_2) in the films.

3. Devices Results and Analysis

Taking into a count the set of results obtained during the evaluation of the films processed, we centred our attention in evaluating the devices performances by using tin oxide channel layers processed close to the extremes of the O_{pp} window in which a clear p-type behaviour was observed after annealing, respectively for O_{pp} = 3.0% (≈1.57 sccm) and O_{pp} = 3.6% (≈1.78 sccm). To reduce the channel conductance, thus allowing a better modulation of the same, we reduced the channel layer thickness to values around 12 nm. A batch of more than 40 devices were evaluated and the devices performances varied within a standard deviation of about ± 7% from the average.

3.1. Devices Structure, Geometry, Fabrication and Characterization Conditions

Bottom gate TFTs were fabricated on glass substrates coated with 150 nm thick layer of sputtered ITO and a 220 nm thick layer of aluminium-titanium oxide (ATO). SnO_x channel layer (width/length = 50 nm/50 nm and 12 nm thick) was deposited over this coating by r.f. magnetron sputtering at RT, using the same process conditions as reported before. Drain and source electrodes were based on Ni/Au (9 nm/60 nm) stack layers deposited by electron beam evaporation. After deposition, the devices were annealed in air up to 200 °C for 30 minutes.

Figure 4a–d show the SEM and AFM images of the surface morphology of the TFT channel layer 12 nm thick. The data depicted show that the shape of the microstructure of the grains obtained are similar, slightly increasing as O_{pp} increases, while the roughness decreases, respectively from 4.6 nm to 4.3 nm.

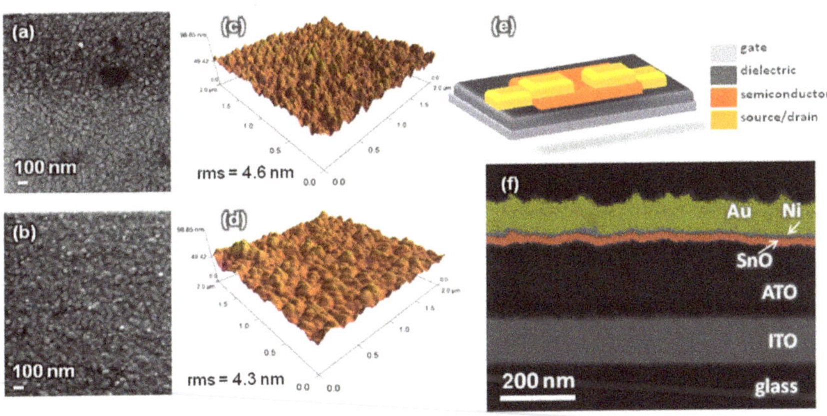

Figure 4. After annealing (60 mn) scanning electron microscope (SEM) images of SnO_x films processed at O_{pp} = 3.0% (**a**) O_{pp} = 3.6% (**b**) 3.6%, respectively. (**c**) and (**d**) show the corresponding AFM images. (**e**) Schematic illustration of SnO_x TFT structures fabricated. (**f**) Cross-sectional SEM image of the fabricated TFT.

Figure 4e is a schematic of the SnO_x TFT with a bottom gate configuration. Figure 4f shows the cross-section SEM image of the TFT prepared at 3.0% O_{pp}, revealing a perfect step coverage of the deposited layer (channel and drain/source contacts), highly compact, uniform and homogeneous, without visible defects.

TFT electrical characterization was performed with an Agilent 4155C semiconductor parameter analyzer (Santa Clara, CA, USA) and a Cascade Microtech M150 microprobe station (Livermore, CA,

USA) inside a dark box at ambient atmosphere. By doing so, we avoid problems that may arise from persistent photoconductive effects [65].

3.2. Capacitance Measurements Data and Analysis

In order to better understand the role of the interfaces as well as of the dielectric on the TFT performances, CV measurements were directly performed on the TFTs by performing measurements between the gate electrode and drain and source short-circuited. The data were interpreted using an electrical model consisting of a contact resistance R_C (the same value for drain and source regions) in series with a combination of two parallel RC resonators as shown in the sketch of Figure 5a. One of the resonators represents the semiconductor channel capacitance (C_S) that varies dynamically, depending on the extension of the accumulation/depletion layer. The other component is the interface trap capacitance (C_{it}) in series with the corresponding associated interface resistance R_{it}, both depending on the interface defects given by:

$$C_{it} = \frac{\rho(V)}{\frac{dV}{dx}} = \varepsilon\varepsilon_0 \frac{\rho(V)}{Q} = \frac{q^2 D_{it}}{1+\omega^2\tau_{it}^2} \tag{3}$$

where D_{it} is the interface trap density, $\tau_{it} = R_{it} \times C_{it}$ is the trap response time. Finally, we have in series the insulator geometric capacitance (C_{ox}). As the frequency tends to a steady state condition, we have almost C_{it} in parallel with C_s and the resulting capacitance in series with C_{ox}. Therefore, for non-perfect semiconductors (basically, the amorphous ones), by using a frequency modulation less than the relaxation frequency of the semiconductor ($f_r = \frac{1}{2\pi R_B C_s} = \frac{1}{2\pi\rho_B\varepsilon_S\varepsilon_0}$, where R_B is the channel bulk resistance), we expect that capacitance will be influenced by the number and nature of the interface defects, their length and depth extension which determines the way in which the structure responds to the electrical stimulus in the low frequency regime. On the other hand, at very high frequencies, C_s dominates and $C_{it} \cong 0$, being now relevant the role of R_{it} in parallel with C_{ox}. In this case the capacitance-voltage curves can be distorted due to charging effects, leading to a decrease or even to a not well-defined flat capacitance maximum (C_{max}), as observed in the normalized C-V plots of Figure 5b for different frequencies and on the behaviour of C_{max} with the frequency, depicted in Figure 5c. Figure 5b also shows the transfer characteristics of the TFT under analysis, where the hysteresis behaviour recorded is visible, which we associate with the role of interface defects.

Figure 5c shows the dependence of the maximum capacitance recorded on the frequency. There, we notice that at frequencies below 100 Hz the capacitance tends to reach flat behaviour while the hysteresis voltage shift is enhanced, following a similar trend as that of the I-V TFT transfer characteristics. We associate this behaviour with interface localized states that respond at frequencies below f_r.

The analysis of the C-V plots shows that the minimum capacitance is not fully flat. This behaviour is attributed to the small thickness of the semiconductor ($\leq$ 12 nm), which limits the extension of the depletion region. As the possible maximum width of depletion region is 12 nm (the semiconductor thickness), this means that the minimum capacitance of the system will not be much lower than the total capacitance. We also estimate the flat band voltage shift and the oxide charge density.

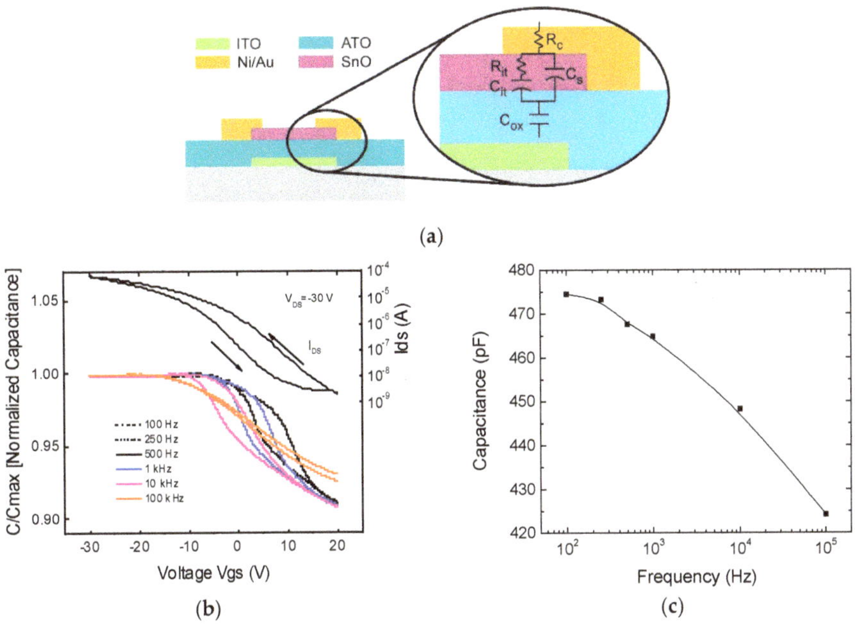

Figure 5. (**a**) Cross section sketch of the TFT, where t is also shown the equivalent electrical circuit based on RC resonators. (**b**) Normalized C-V plot taken for different frequencies (below) and the IV transfer characteristic of the TFT (above); (**c**) Dependence of the maximum capacitance (C_{max}) achieved on the frequency used.

Figure 6 shows the dependence of these parameters on the frequency, for the upward and downwards sweeps, in order to take into account the hysteresis observed. For the set of calculations performed, it was considered that the level of acceptor concentration (N_a) was in the range of 10^{17} cm^{-3} with a good work function match between the gate electrode and the semiconductor.

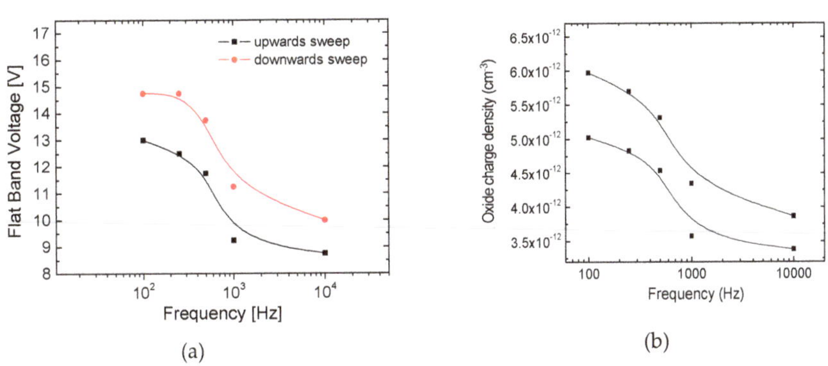

Figure 6. (**a**) Flat band Voltage dependence on the frequency; (**b**) Oxide charge density as a function of the frequency. The upper curves correspond to the upwards sweep while the lower curves to the downwards sweep (see Figure 5b).

Overall the data depicted show an enhancement on the flat band voltage and on the oxide charge density as the frequency decreases.

3.3. TFT Electrical Data and Analysis

Figures 7 and 8 show the TFT electrical characteristics of the two extreme cases evaluated, as processed, and after 50 days as a way to determine their stability (devices ageing effects).

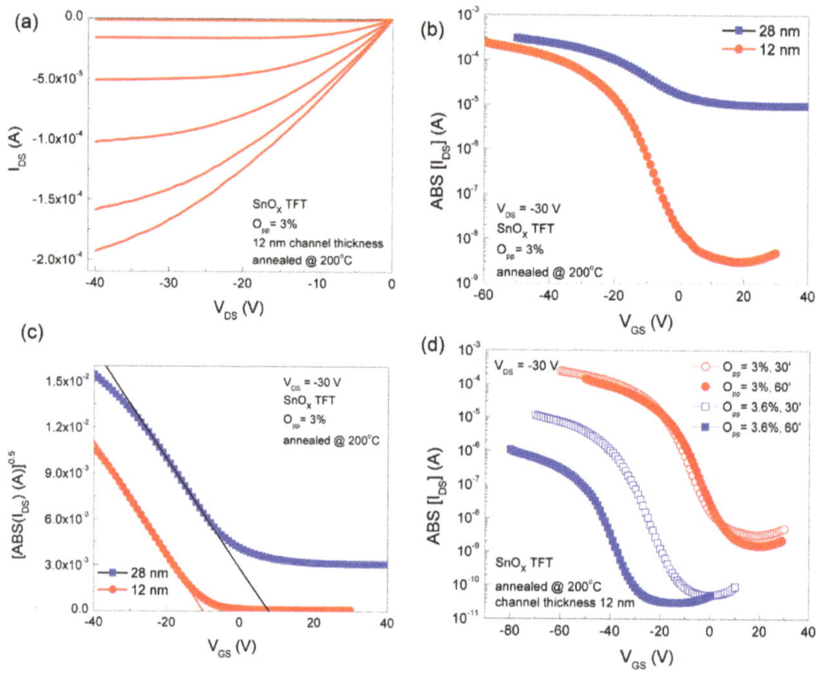

Figure 7. (**a**) Output characteristics of SnO_x p-channel TFT. The gate voltage is varied from 0 V to −50 V in −10 V steps. (**b**) Transfer characteristics of SnO_x p-channel TFT with different channel thickness. (**c**) Plot of square root of drain current, at V_{DS} = −30 V, for evaluating saturation mobility and threshold voltage. (**d**) Transfer characteristics of SnO_x p-channel TFT (channel thickness 12 nm) for different annealing time.

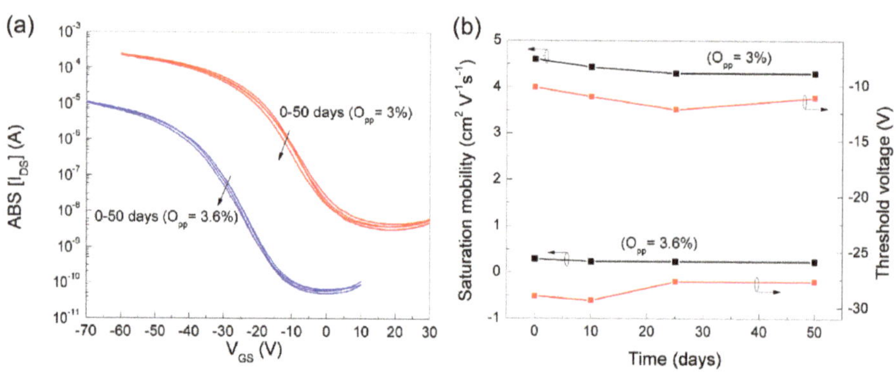

Figure 8. Transfer characteristics of SnO_x p-channel TFT measured at different days after device fabrication (**a**) and corresponding variations in saturation mobility and threshold voltage (**b**).

Figure 7a shows the output characteristics (I_{DS}-V_{DS}) of such TFT produced at O_{pp} = 3.0%. The gate voltage was varied from 0 V to −50 V in −10 V steps. Very small I_{DS} at zero gate voltage indicates an almost closed channel. On increasing the gate voltage to higher negative values, holes accumulated at the channel-insulator interface forms a conduction path between source and drain. These TFTs exhibit hard saturation at large V_{DS}, which is similar to pinch off in the usual field-effect transistors, revealing the high quality of the devices fabricated.

Figure 7b shows the transfer characteristics of the same TFT, as above, but now using two different thicknesses for the channel layer, 12 nm and 28 nm respectively. The data depicted reveal that by decreasing the channel thickness the On/Off ratio improves from ~10^2 to 7×10^4, while the subthreshold swing voltage (the slop of the transfer curve when the current goes from the Off state to the On state) is reduced by more than a factor of 3. This allows us to select 12 nm as the best device thickness as stated before.

Figure 7c shows the plot of the square root of the absolute value of the drain current, at $V_{DS} = -30$ V as a function of V_{GS}, for the same samples depicted in Figure 7b, for evaluating saturation mobility and threshold voltage, respectively. The data depicted show that irrespective of the channel layer thickness, the saturation mobility achieved is around 4.6 $cm^2\ V^{-1}\ s^{-1}$, while the threshold voltage is shifted from 8.1 V to -10 V on reducing the channel thickness from 28 nm to 12 nm, respectively. Negative threshold voltage indicates the enhancement mode operation of the p-channel TFTs, as required to keep the channel closed when no voltage is applied to the gate.

Figure 7d shows the dependence of the absolute value of I_{DS} [ABS(I_{DS})] as a function of V_{GS} for devices fabricated using O_{pp} equal to 3% and 3.6%, respectively and heat treated at 200 °C during 30 mn and 60 mn, respectively. Table 4 shows the TFT parameters extracted from Figure 7d, where the dominant phase composition of the films processed is also shown, as revealed by the difractograms in Figure 1. The data show that the most stable devices with the best device performances are achieved when using an O_{pp} of about 3% and an annealing time of around 30 minutes, where the main dominant phase is just attributed to α-SnO. On the other hand, the films processed at $O_{pp} \geq 3.6\%$ reveal a growing role of the β-SnO phase and so increase the ambipolar behaviour of the films processed.

Table 4. Electrical properties of SnO_x p-channel TFTs for different annealing time (channel thickness 12 nm). After annealing all structures are polycrystalline.

Annealing Conditions	Oxygen Partial Pressure (%)	Field Effect Mobility ($cm^2\ V^{-1}\ s^{-1}$)	Saturation Mobility ($cm^2\ V^{-1}\ s^{-1}$)	Threshold Voltage (V)	On-Off Ratio	Dominant Composition Phase
200 °C, 30 min	3.0	3.3	4.6	−10	7×10^4	α-SnO
200 °C, 60 min	3.0	2.2	2.6	−7.2	9×10^4	α-SnO
200 °C, 30 min	3.6	0.16	0.28	−29	2×10^5	α-SnO + β-SnO
200 °C, 60 min	3.6	0.02	0.03	−38.6	3×10^4	α-SnO + β-SnO

Figure 8 shows the transfer curves of TFTs fabricated using O_{pp} = 3% and O_{pp} = 3.6%, measured along consecutive 50 days after their fabrication. The data show that that the TFTs are very stable in operation, without any significant variation on On/Off ratio, threshold voltage and mobility.

4. Discussion of the Results

The analysis of the structure, composition and morphology of the films processed, together with the evaluation of their electrical and optical performances, allow us to select the best process conditions to produce high stable and reliable p-type tin oxide films by RFMS at RT, surpassing the existing state of the art knowledge, as shown in Table 1. The data shown in Figure 1 reveal a structure in which the metallic tin dominates over the SnO phase, as also confirmed by Mössbauer spectroscopy data. This suggests that in the α-SnO phase, the Sn^{2+} oxide present is amorphous. After annealing in air, the metallic tin (β-Sn) is oxidized leading to the SnO_x ($1 < x < 2$) phase formation, where now the Sn^{2+} oxide present is crystalline, with a strong α-SnO phase for which the minimum heat treatment time required is of about 30 min. To reach this goal O_{pp} of $\geq 2.8\%$ is needed which corresponds with the use of an oxygen gas flow above 1.46 sccm.

Below an oxygen gas flow of 1.46 sccm, films are metallic in nature and adhesion to the glass substrate was very poor.

On the other hand, if O_{pp} > 3.8%, which corresponds with the use of an oxygen gas flow above 1.89 sccm, the films are highly resistive (>10^8 Ω·cm), decreasing substantially after annealing, with SnO_2 being the dominant phase. In these conditions, films exhibit an n-type conduction behaviour. This clearly favours the ambipolar behaviour, as broadly observed [38,42,51,53].

Since we used a metallic tin target for sputtering, very low O_{pp} was not sufficient to oxidize the film. The experimental data show that a stable SnO phase is only obtained if used during the deposition process of an oxygen gas flow in the range of 1.46–1.89 sccm (2.8% < O_{pp} <3.8%.), after heat treatment at 200 °C for 30 minutes. This corresponds to producing films with a resistivity in the range of 30–40 Ω.cm (see Figure 2). On the other hand, the use of high oxygen gas flow (more than 1.89 sccm) favours SnO_2 formation. That is, we can use the same material and, by proper control of the process parameters such as O_{pp} and gas flow, turn it into an n-type or p-type semiconductor [51,53].

The optical data shows that the transparency in the visible range for films as deposited is below 20%, increasing as O_{pp} increases. Moreover, the heat treatment enhances the optical transparency by about 3 times, proving that under these conditions the oxidation phase dominates over the metallic one (see Figure 3). The estimated bandgap value for the films deposited using oxygen flow in the range of 1.46–1.89 sccm was around 2.6 eV (see Figure 3). This agrees with the reported direct bandgap value of 2.7 eV, associated with the SnO dominant phase.

The study described above allowed us to select the best process conditions to grow p-type tin oxide films as being those for which O_{pp} = 3%. Moreover, allow us also to observe that the thickness of the films plays a relevant role in determining the electrical modulating behaviour of the channel, as also observed in other works [48]. This led us to select the proper thickness of the channel layer as being 12 nm, as shown in Figure 4.

The analysis of the capacitance measurements realized on the TFT (pad contact area of about 100 μm × 100 μm) shows that the capacitance reaches a flat maximum for frequencies below 100 Hz, which corresponds to a value of about 23.8 nF/μm^2. This allowed us to infer the saturation mobility [15]. Here, the reduced channel thickness ($\leq$ 12 nm), limits the extension in which the channel is depleted and also its modulation extension. The hysteresis observed is attributed to the role of interface localized states in responding to electrical stimulus below the relaxation frequency f_r. Moreover, from these measurements we could estimate the flat band voltage shift and the oxide charge density (see Figure 6) as being in the range from 13–15 V and 5×10^{12}–6×10^{12} C/cm^3, for the low frequency regime, shifting towards 9–10 V and 3.5×10^{12}–4×10^{12} C/cm^3, respectively for the high frequency regime, where shallow states do not respond to the electrical stimulus [15].

The analysis of the electrical characteristics of the TFT depicted in Figures 7 and 8 show that p-channel oxide TFTs manufactured at RT, followed by intentional heat treatment at 200 °C, using a narrow process window associated with the oxygen gas flow used are highly reproducible, reliable and stable. In addition, they all work in the enhancement mode, exhibiting very good electronic performance, translated by high mobility and On/Off ratio, together with a relative low threshold voltage. The annealing temperature used is suitable for fabricating these devices on plastic or even on paper substrates [7], which cannot be the case when using heat treatment much above 200 °C [29–32]. The data depicted clearly show that the films developed throughout this study are among the best ever produced, as can be seen in Tables 1 and 2, irrespective of the process technique used. Moreover, the very high On current of these devices makes them ideal for fabricating active matrix OLED driving circuits, where a TFT must supply sufficient hole current to the anode of the OLED. Use of an n-channel TFT results in a voltage drop over the OLED, which affects the drain current of the TFT. However, it does not affect the drain current of a p-channel TFT in the saturation mode.

5. Conclusions

In summary, we have fabricated highly reproducible high-performance p-channel oxide TFTs on glass substrates using an SnO_x channel layer, for which the proper process conditions were selected. The SnO phase was identified and quantified by two independent techniques, XRD and Mossbauer spectroscopy, corroborating the p-type oxide semiconductor behaviour obtained, for films grown at low O_{pp} (around 3%), where the dominant phase is due to α-SnO. The TFTs fabricated show typical saturation mobility of 4.6 $cm^2\ V^{-1}\ s^{-1}$ and On-Off ratio > 7×10^4, with threshold voltages of about -10 V, indicating that the devices work in the enhancement mode. This is one of the best stable electrical performances achieved so far for p-type oxide TFTs processed at RT and heat treated at temperatures around 200 °C, as the survey conducted concerning the present state of the art in producing p-type TFT based on tin oxide shows (see Tables 1 and 2). This will enable the fabrication of fully transparent CMOS either on rigid or flexible substrates, associated with all the advantages offered by transparent/oxide electronics.

Author Contributions: Conceptualization, E.F. and K.J.S.; methodology, R.M.; Structure, morphology, electrical and optical characterization, R.B., P.B.; Mössbauer spectra analysis J.C.W.; validation, E.F., P.B., R.B., K.J.S.; formal analysis, R.B., P.B., L.P.; resources, E.F., R.M.; data curation, P.B., R.B.; writing—original draft preparation, K.J.S.; R.M.; writing—review and editing; overview on the present state of art on p-type TFT SnO based, E.C., R.M.; visualization, E.F., R.M.; supervision, E.F.; project administration, E.F.; funding acquisition, E.F.

Funding: This work was partially supported by the European Commission under project INVISIBLE (Advanced Grant from ERC nº 228144); BET-EU (H2020-TWINN-2015, GA 692373); TREND (ERC-StG-2016, grant GA 716510); 1D Neon (H2020-NMP-2015-IA, grant 685758-21D). We also acknowledge the FEDER funds through the COMPETE 2020 Program under the project UID/CTM/50025/2013). The authors K. J. Saji thanks Portuguese Ministry of Science and Technology (FCT-MCTES) for offering research grants through the fellowships SFRH/BPD/62942/2009.

Acknowledgments: The authors would like to thank Joana Pinto for the support given in measuring the XRD of the films processed as well as the support given by the CENIMAT technical and administrative staff throughout the present study.

Conflicts of Interest: The authors declare no conflict of interest.

References

1. Klasens, H.A.; Koelmans, H. A Tin oxide Field-Effect Transistor. *Solid-State Electron.* **1964**, *7*, 701–702. [CrossRef]
2. Boesen, G.F.; Jacobs, J.E. ZnO Field Effect Transistor. *Proc. IEEE* **1968**, *56*, 2094. [CrossRef]
3. Nomura, K.; Ohta, H.; Takagi, A.; Kamiya, T.; Hirano, M.; Hosono, H. Room-temperature fabrication of transparent flexible thin-film transistors using amorphous oxide semiconductors. *Nature* **2004**, *435*, 488. [CrossRef] [PubMed]
4. Hoffman, R.L.; Norris, B.J.; Wager, J.F. ZnO-based transparent thin-film transistors. *Appl. Phys. Lett.* **2003**, *82*, 733. [CrossRef]
5. Carcia, P.F.; McLean, R.S.; Reilly, M.H.; Nunes, G. Transparent ZnO thin-film transistor fabricated by rf magnetron sputtering. *Appl. Phys. Lett.* **2003**, *82*, 1117.
6. Fortunato, E.; Barquinha, P.; Pimental, A.; Goncalves, A.; Marques, A.; Pereira, L.; Martins, R. Fully transparent ZnO thin-film transistor produced at room temperature. *Adv. Mater.* **2005**, *17*, 590. [CrossRef]
7. Martins, R.; Ferreira, I.; Fortunato, E. Electronics with and on paper. *Phys. Status Solidi-RRL* **2011**, *5*, 332–335. [CrossRef]
8. Fortunato, E.; Martins, R. Where science fiction meets reality? With oxide semiconductors! *Phys. Status Solidi-RRL* **2011**, *5*, 336. [CrossRef]
9. Senanayak, S.P.; Sangwan, V.K.; McMorrow, J.J.; Everaerts, K.; Chen, Z.H.; Facchetti, A.; Hersam, M.C.; Marks, T.J.; Narayan, K.S. Self-Assembled Photochromic Molecular Dipoles for High-Performance Polymer Thin-Film Transistors. *ACS Appl. Mater. Interfaces* **2018**, *10*, 21492. [CrossRef] [PubMed]
10. Kim, J.Y.; Kim, J.W.; Lee, E.K.; Park, J.I.; Lee, B.L.; Kwon, Y.N.; Byun, S.; Jung, M.S.; Kim, J.J. Cross-linked poly(hydroxy imide) gate-insulating materials for low-temperature processing of organic thin-film transistors. *J. Mater. C* **2018**, *6*, 13359–13366. [CrossRef]

11. Street, R.A. Thin-Film Transistors. *Adv. Mater.* **2009**, *21*, 2007. [CrossRef]
12. Dimitrakopoulos, C.D.; Malenfant, P.R.L. Organic thin film transistors for large area electronics. *Adv. Mater.* **2002**, *14*, 99. [CrossRef]
13. Chen, F.Z.; Jiang, Y.; Sui, Y.; Zhang, J.D.; Tian, H.K.; Han, Y.; Deng, Y.F.; Hu, W.P.; Geng, Y.H. Donor-Acceptor Conjugated Polymers Based on Bisisoindigo: Energy Level Modulation toward Unipolar n-Type Semiconductors. *Macromolecules* **2018**, *51*, 8652–8661. [CrossRef]
14. Kumagai, S.; Nakano, M.; Takimiya, K.; Takeya, J. Solution-crystallized n-type organic thin-film transistors: An impact of branched alkyl chain on high electron mobility and thermal durability. *Org. Electron.* **2018**, *62*, 548–553. [CrossRef]
15. Fortunato, E.; Barquinha, P.; Martins, R. Oxide Semiconductor Thin-Film Transistors: A Review of Recent Advances. *Adv. Mater.* **2012**, *24*, 2945–2986. [CrossRef] [PubMed]
16. Fernandes, C.; Santa, A.; Santos, A.; Bahubalindruni, P.; Deuermeier, J.; Martins, R.; Fortunato, E.; Barquinha, P. A Sustainable Approach to Flexible Electronics with Zinc-Tin Oxide Thin-Film Transistors. *Adv. Elecron. Mater.* **2018**, *4*, 1800032. [CrossRef]
17. Fortin, E.; Weichman, F.L. Hall effect and electrical conductivity of Cu_2O monocrystals. *Can. J. Phys.* **1966**, *44*, 1551. [CrossRef]
18. Matsuzaki, K.; Nomura, K.; Yanagi, H.; Kamiya, T.; Hirano, M.; Hosono, H. Epitaxial growth of high mobility Cu(2)O thin films and application to p-channel thin film transistor. *Appl. Phys. Lett.* **2008**, *93*, 202107. [CrossRef]
19. Deuermeier, J.; Fortunato, E.; Martins, R.; Klein, A. Energy band alignment at the nanoscale. *Appl. Phys. Lett.* **2017**, *110*, 051603. [CrossRef]
20. Liu, A.; Nie, S.B.; Liu, G.X.; Zhu, H.H.; Zhu, C.D.; Shin, B.; Fortunato, E.; Martins, R.; Shan, F. In situ one-step synthesis of p-type copper oxide for low-temperature, solution-processed thin-film transistors. *J. Master. Chem. C* **2017**, *5*, 2524. [CrossRef]
21. Liu, A.; Zhu, H.H.; Guo, Z.D.; Meng, Y.; Liu, G.X.; Fortunato, E.; Martins, R.; Shan, F.K. Solution Combustion Synthesis: Low-Temperature Processing for p-Type Cu:NiO Thin Films for Transparent Electronics. *Adv. Mater.* **2017**, *29*, 1701599. [CrossRef] [PubMed]
22. Shan, F.K.; Liu, A.; Zhu, H.H.; Kong, W.J.; Liu, J.Q.; Shin, B.C.; Fortunato, E.; Martins, R.; Liu, G.X. High-mobility p-type NiOx thin-film transistors processed at low temperatures with Al2O3 high-k dielectric. *J. Mater. Chem. C* **2016**, *4*, 9438–9444. [CrossRef]
23. Stedile, F.C.; Debarros, B.A.S.; Leite, C.B.; Freire, F.L.; Baumvol, I.J.R.; Schreiner, W.H. Characterization of Tin Oxide Thin Films depopsited by reactive sputtering. *Thin Solid Films* **1989**, *170*, 285–291. [CrossRef]
24. Togo, A.; Oba, F.; Tanaka, I.; Tatsumi, K. First-principles calculations of native defects in tin monoxide. *Phys. Rev. B* **2006**, *74*, 195128. [CrossRef]
25. Allen, J.P.; Scanlon, D.O.; Parker, S.C.; Watson, G.W. Tin Monoxide: Structural Prediction from First Principles Calculations with van der Waals Corrections. *J. Phys. Chem. C* **2011**, *115*, 19916. [CrossRef]
26. Pan, X.Q.; Fu, L. Tin oxide thin films grown on the ((1)over-bar012) sapphire substrate. *J. Electrocer.* **2001**, *7*, 35. [CrossRef]
27. Batzill, M.; Diebold, U. The surface and materials science of tin oxide. *Prog. Surf. Sci.* **2005**, *79*, 47. [CrossRef]
28. Fortunato, E.; Barros, R.; Barquinha, P.; Figueiredo, V.; Park, S.H.; Hwang, C.S.; Martins, R. Transparent p-type SnOx thin film transistors produced by reactive rf magnetron sputtering followed by low temperature annealing. *Appl. Phys. Lett.* **2010**, *97*, 052105. [CrossRef]
29. Caraveo-Frescas, J.A.; Jesus, A.; Nayak, P.K.; Al-Jawhari, H.A.; Granato, D.B.; Schwingenschlogl, U.; Alshareeft, H.N. Record Mobility in Transparent p-Type Tin Monoxide Films and Devices by Phase Engineering. *ACS Nano* **2013**, *7*, 5160. [CrossRef] [PubMed]
30. Zhong, C.W.; Lin, H.C.; Liu, K.C.; Huang, T.Y. Improving Electrical Performances of p-Type SnO Thin-Film Transistors Using Double-Gated Structure. *IEEE Electron. Device Lett.* **2015**, *36*, 1053–1055. [CrossRef]
31. Shih, C.W.; Chin, A.; Lu, C.F.; Su, W.F. Remarkably High Hole Mobility Metal-Oxide Thin-Film Transistors. *Sci. Rep.* **2018**, *8*, 889. [CrossRef] [PubMed]
32. Yabuta, H.; Kaji, N.; Hayashi, R.; Kumomi, H.; Nomura, K.; Kamiya, T.; Hirano, M.; Hosono, H. Sputtering formation of p-type SnO thin-film transistors on glass toward oxide complimentary circuits. *Appl. Phys. Lett.* **2010**, *97*, 072111. [CrossRef]

33. Martins, R.; Nathan, A.; Barros, R.; Pereira, L.; Barquinha, P.; Correia, N.; Costa, R.; Ahnood, A.; I Ferreira, I.; Fortunato, E. Complementary Metal Oxide Semiconductor Technology with and on Paper. *Adv. Mater.* **2011**, *23*, 4491. [CrossRef] [PubMed]
34. Hsu, P.-C.; Chen, W.-C.; Tsai, Y.-T.; Kung, Y.-C.; Chang, C.-H.; Hsu, C.-J.; Wu, C.-C.; Hsieh, H.-H. Fabrication of p-Type SnO Thin-Film Transistors by Sputtering with Practical Metal Electrodes. *Jpn. J. Appl. Phys.* **2013**, *52*, 05DC07. [CrossRef]
35. Chiu, I.C.; Li, Y.-S.; Tu, M.-S.; Cheng, I.C. Complementary Oxide–Semiconductor-Based Circuits With n-Channel ZnO and p-Channel SnO Thin-Film Transistors. *IEEE Electron Device Lett.* **2014**, *35*, 1263. [CrossRef]
36. Myeonghun, U.; Han, Y.; Song, S.; Cho, I.; Lee, J.; Kwon, H.-I. High Performance p-type SnO thin-film Transistor with SiOx Gate Insulator Deposited by Low-Temperature PECVD Method. *J. Semicond. Technol. Sci.* **2014**, *14*, 666.
37. Caraveo-Frescas, J.A.; Khan, M.A.; Alshareef, H.N. Polymer ferroelectric field-effect memory device with SnO channel layer exhibits record hole mobility. *Sci. Rep.* **2014**, *4*, 1. [CrossRef] [PubMed]
38. Luo, H.; Liang, L.Y.; Liu, Q.; Cao, H.T. Magnetron-Sputtered SnO Thin Films for p-Type and Ambipolar TFT Applications. *ECS J. Solid State Sci. Technol.* **2014**, *3*, Q3091. [CrossRef]
39. Khan, M.A.; Caraveo-Frescas, J.A.; Alshareef, H.N. Hybrid dual gate ferroelectric memory for multilevel information storage. *Org. Electron.* **2015**, *16*, 9. [CrossRef]
40. Han, Y.; Choi, Y.; Jeong, H.; Kwon, H.I. Investigation of Intrinsic Electrical Characteristics and Contact Effects in p-Type Tin Monoxide Thin-Film Transistors Using Gated-Four-Probe Measurements. *J. Nanosci. Nanotechnol.* **2015**, *15*, 7582. [CrossRef] [PubMed]
41. Han, Y.J.; Choi, Y.J.; Jeong, C.Y.; Lee, D.; Song, S.H.; Kwon, H.I. Environment-Dependent Bias Stress Stability of P-Type SnO Thin-Film Transistors. *IEEE Electron Device Lett.* **2015**, *36*, 466. [CrossRef]
42. Saji, K.J.; Mary, A.P.R. Tin Oxide Based P and N-Type Thin Film Transistors Developed by RF Sputtering. *ECS J. Solid State Sci. Technol.* **2015**, *4*, Q101. [CrossRef]
43. Hsu, P.C.; Tsai, S.P.; Chang, C.H.; Hsu, C.J.; Chen, W.C.; Hsieh, H.H.; Wu, C.C. Preparation of p-type SnO thin films and transistors by sputtering with robust Sn/SnO2 mixed target in hydrogen-containing atmosphere. *Thin Solid Films* **2015**, *585*, 50. [CrossRef]
44. Zhong, C.-W.; Lin, H.-C.; Tsai, J.-R.; Liu, J.-R.; Huang, T.-Y. Impact of gate dielectrics and oxygen annealing on tin-oxide thin-film transistors. *Jpn. J. Appl. Phys.* **2016**, *55*, 04EG02. [CrossRef]
45. Bae, S.; Kwon, S.-H.; Jeong, H.; Kwon, H. Demonstration of high-performance p-type tin oxide thin-film transistors using argon-plasma surface treatments. *Semicond. Sci. Technol.* **2017**, *32*, 075006. [CrossRef]
46. Yang, V.; Wang, Y.; Li, Y.; Yuan, Y.; Hu, Z.; Ma, P.; Zhou, L.; Wang, Q.; Song, A.; Xin, Q. Highly Optimized Complementary Inverters Based on p-SnO and n-InGaZnO With High Uniformity. *IEEE Electron Device Lett.* **2018**, *39*, 516. [CrossRef]
47. Guan, X.; Wang, Z.; Hota, M.K.; Alshareef, H.N.; Wu, T. P-Type SnO Thin Film Phototransistor with Perovskite-Mediated Photogating. *Adv. Electron. Mater.* **2019**, *1*, 1800538. [CrossRef]
48. Qu, Y.; Yang, J.; Li, Y.; Zhang, J.; Wang, Q.; Song, A.; Xin, Q. Organic and inorganic passivation of p-type SnO thin-film transistors with different active layer thicknesses. *Semicond. Sci. Technol.* **2018**, *33*, 075001. [CrossRef]
49. Li, Y.; Zhang, J.; Yang, J.; Yuan, Y.; Hu, Z.; Lin, Z.; Song, A.; Xin, Q. Complementary Integrated Circuits Based on n-Type and p-Type Oxide Semiconductors for Applications Beyond Flat-Panel Displays. *IEEE Trans. Electron Devices* **2019**, *66*, 950. [CrossRef]
50. Ogo, Y.; Hiramatsu, H.; Nomura, K.; Yanagi, H.; Kamiya, T.; Hirano, M.; Hosono, H. p-channel thin-film transistor using p-type oxide semiconductor, SnO. *Appl. Phys. Lett.* **2008**, *93*, 032113.
51. Nomura, K.; Kamiya, T.; Hosono, H. Ambipolar Oxide Thin-Film Transistor. *Adv. Mater.* **2011**, *23*, 3431. [CrossRef] [PubMed]
52. Hsu, P.-C.; Wu, C.-C.; Hiramatsu, H.; Kamiya, T.; Hosono, H. Film Texture, Hole Transport and Field-Effect Mobility in Polycrystalline SnO Thin Films on Glass. *ECS J. Solid State Sci. Technol.* **2014**, *3*, Q3040. [CrossRef]
53. Liang, L.; Cao, H. Ambipolar SnO thin-film transistors and inverters. *ECS Trans.* **2012**, *50*, 289. [CrossRef]
54. Hung, M.P.; Genoe, J.; Heremans, P.; Steudel, S. Off-current reduction in p-type SnO thin film transistors. *Appl. Phys. Lett.* **2018**, *112*, 2635021.

55. Chen, P.-C.; Chiu, Y.-C.; Zheng, Z.-W.; Cheng, C.-H.; Wu, Y.-H. P-type tin-oxide thin film transistors for blue-light detection application. *Phys. Status Solidi-Rapid Res. Lett.* **2016**, *10*, 919. [CrossRef]
56. Chen, P.-C.; Chiu, Y.-C.; Zheng, Z.-W.; Cheng, C.-H.; Wu, Y.-H. Influence of plasma fluorination on p-type channel tin-oxide thin film transistors. *J. Alloys Compd.* **2017**, *707*, 162. [CrossRef]
57. Kim, S.K.H.; Baek, I.-H.; Kim, D.H.; Pyeon, J.J.; Chung, T.-M.; Baek, S.-H.; Kim, J.-S.; Han, J.H.; Kim, S.K. Fabrication of high-performance p-type thin film transistors using atomic-layer-deposited SnO films. *J. Mater. Chem. C* **2017**, *5*, 3139. [CrossRef]
58. Okamura, K.; Nasr, B.; Brand, R.A.; Hahn, H. Solution-processed oxide semiconductor SnO in p-channel thin-film transistors. *J. Mater. Chem.* **2012**, *22*, 4607. [CrossRef]
59. Shenoy, G.K.; Wagner, F.E. *Mössbauer Isomer Shifts*; North Holland Publ. Co.: Amsterdam, The Netherlands, 1978.
60. Blanca, E.; Svane, A.; Christensen, N.E.; Rodriguez, C.O.; Cappannini, O.M.; Moreno, M.S. Calculated Static and dynamic Properties of beta-Sn and Sn-O compounds. *Phys. Rev. B* **1993**, *48*, 15712. [CrossRef]
61. Conte, D.E.; Aboulaich, A.; Robert, F.; Olivier-Fourcade, J.; Jumas, J.C.; Jordy, C.; Willmann, P. Sn-x[BPO4](1-x) composites as negative electrodes for lithium ion cells: Comparison with amorphous SnB0.6P0.4O2.9 and effect of composition. *J. Solid State Chem.* **2010**, *183*, 65. [CrossRef]
62. Herber, R.H. Mössbauer Lattice temperature of Tetragonal (P4NMM) SnO. *Phys. Rev. B* **1983**, *27*, 4013. [CrossRef]
63. Moreno, M.S.; Mercader, R.C. Mössbauer study of SnO Lattice dynamics. *Phys. Rev. B* **1994**, *50*, 9875. [CrossRef]
64. Hosono, H.; Ogo, Y.; Yanagi, H.; Kamiya, T. Bipolar Conduction in SnO Thin Films. *Electrochem. Solid State Lett.* **2011**, *14*, II13. [CrossRef]
65. Jeon, S.; Ahn, S.E.; Song, I.; Kim, C.J.; Chung, U.I.; Lee, E.; Yoo, I.; Nathan, A.; Lee, S.; Robertson, J.; et al. Gated three-terminal device architecture to eliminate persistent photoconductivity in oxide semiconductor photosensor arrays. *Nat. Mater.* **2012**, *11*, 301–305. [CrossRef] [PubMed]

nanomaterials

MDPI

Article

Fully Printed Zinc Oxide Electrolyte-Gated Transistors on Paper

José Tiago Carvalho [1], Viorel Dubceac [1], Paul Grey [1], Inês Cunha [1], Elvira Fortunato [1], Rodrigo Martins [1], Andre Clausner [2], Ehrenfried Zschech [2] and Luís Pereira [1,*]

1 CENIMAT/I3N, Departamento de Ciência dos Materiais, Faculdade de Ciências e Tecnologia, FCT, Universidade Nova de Lisboa and CEMOP-UNINOVA, Campus da Caparica, 2829-516 Caparica, Portugal; jt.carvalho@campus.fct.unl.pt (J.T.C.); v.dubceac@campus.fct.unl.pt (V.D.); paul16_grey@yahoo.de (P.G.); i.cunha@campus.fct.unl.pt (I.C.); emf@fct.unl.pt (E.F.); rfpm@fct.unl.pt (R.M.)

2 Fraunhofer Institute for Ceramic Technologies and Systems (IKTS), 01109 Dresden, Germany; andre.clausner@ikts.fraunhofer.de (A.C.); ehrenfried.zschech@ikts.fraunhofer.de (E.Z.)

* Correspondence: lmnp@fct.unl.pt

Received: 28 December 2018; Accepted: 22 January 2019; Published: 30 January 2019

Abstract: Fully printed and flexible inorganic electrolyte gated transistors (EGTs) on paper with a channel layer based on an interconnected zinc oxide (ZnO) nanoparticle matrix are reported in this work. The required rheological properties and good layer formation after printing are obtained using an eco-friendly binder such as ethyl cellulose (EC) to disperse the ZnO nanoparticles. Fully printed devices on glass substrates using a composite solid polymer electrolyte as gate dielectric exhibit saturation mobility above 5 cm^2 V^{-1} s^{-1} after annealing at 350 °C. Proper optimization of the nanoparticle content in the ink allows for the formation of a ZnO channel layer at a maximum annealing temperature of 150 °C, compatible with paper substrates. These devices show low operation voltages, with a subthreshold slope of 0.21 V dec^{-1}, a turn on voltage of 1.90 V, a saturation mobility of 0.07 cm^2 V^{-1} s^{-1} and an I_{on}/I_{off} ratio of more than three orders of magnitude.

Keywords: zinc oxide; nanoparticles; paper transistors; printed electronics; electrolyte-gated transistors

1. Introduction

Printed electronics (PE) has attracted strong interest among researchers aiming to introduce simpler fabrication techniques for low cost consumer electronics. PE uses well established processes, such as flexo, screen, rotary-screen, inkjet, and off-set printing, and the use of lithography and subtractive etching steps is avoided [1]. These processes offer unique large area processing capability and high throughput on flexible substrates by roll-to-roll (R2R) processes, and it opens a path toward ultra-cost-effective, flexible, and environmentally friendly electronic devices, demanded by modern society standards.

Electrolyte-gated transistors (EGTs) have attracted great interest for flexible electronics and biosensing applications over the last decade [2]. In EGTs, an electrolyte is used as a gate dielectric between the gate electrode and the channel layer. Thus, applying a voltage to the gate electrode causes ion migration in the electrolyte, leading to the formation of an electric double layer (EDL) at the electrolyte/channel interface. This results in large gate capacitance (virtually independent of the thickness) in the order of 1–10 μF cm^{-2}, enabling charge accumulation into the semiconductor at low gate voltages [3]. This makes EGTs very attractive for PE, where thin insulating dielectric layers capable of providing high gate capacitance are difficult to obtain by printing techniques at temperatures compatible with paper substrates [4–7]. Moreover, EGTs are highly sensitive to the ionic species that may contact or exist in the electrolyte. Due to their low operation voltages (<2 V) EGTs are highly desirable for biosensing applications, i.e., to determine and to quantify biological molecules inside aqueous media [8–10].

Solution-processed inorganic semiconductors used in printed EGTs have advantages over organic ones, such as higher intrinsic carrier mobility and stability [11,12]. However, in comparison to organic alternatives, high temperatures are needed to convert the precursors and to synthesize inorganic semiconductors from solution as well as to remove binders or stabilizers from the printed semiconductor ink [13]. These so-called burn-out temperatures could compromise their application on paper substrates. In this context, oxide semiconductor nanoparticle (NP)-based layers are an alternative, avoiding the need of high temperature and long or complex annealing processes to convert a precursor and to form a semiconducting layer.

This study reports the development of fully printed EGTs, based on an interconnected ZnO nanoparticle matrix, arranged in a staggered-top gate architecture. Ethyl cellulose (EC), a cellulose derivative, has been proven to be an indispensable binder in the developed inks, providing the necessary rheological properties for screen printing. Additionally, this eco-friendly biopolymer provides excellent printability and leveling [14] combined with excellent adhesion to paper substrates. Proper formulation of the ZnO printable ink allows one to process these devices at temperatures below 150 °C, totally compatible with paper substrates.

2. Materials and Methods

ZnO ink formulation for screen-printing: The semiconductor ink consists of a dispersion of ZnO NPs in a cellulose derivative, the ethyl cellulose (EC). ZnO nanopowder (Sigma-Aldrich, St. Louis, MO, USA, <100 nm particle size) was dispersed with different concentrations (10 wt % ZnO10 and 40 wt % ZnO40) into a solution of 5 wt % EC (Sigma-Aldrich, St. Louis, MO, USA), viscosity 300 cP, extent of labeling: 48% ethoxyl) in a solvent mixture of toluene/ethanol (80:20 v/v) supplied from Merck (Darmstadt, Germany), ≥99% and Fisher Scientific (Pittsburgh, PA, USA), respectively. The resultant mixture was stirred for 12 h to obtain a well dispersed and homogeneous white viscous solution. Finally, the prepared inks were stored in a refrigerator at 3 °C, due to the high volatility of the components, until use.

Composite solid polymer electrolyte (CSPE) ink formulation for screen-printing: The lithium-based polymer electrolyte ink is composed of a mixture of succinonitrile (Aldrich, St. Louis, MO, USA), an acrylic thermoplastic resin (TB 3003 K, ThreeBond, Saint-Ouen-l'Aumône, France), a lithium salt ($LiClO_4$, Sigma-Aldrich), and titanium dioxide NPs (TiO_2, Rockwood Pigments NA, Inc., Beltsville, MD, USA). All proceeding steps were carried out at 60 °C and relative humidity of 50%.

Electrolyte-gated transistor fabrication and characterization: The fully printed ZnO NP EGTs were produced with a staggered-top gate structure onto glass (Marienfeld) and paper substrates, by a custom-made screen-printing. All materials were used without any further purification. Carbon interdigital source/drain (S/D) electrodes (effective W/L = 39725 μm/330 μm ≈ 120) were printed by a polyester mesh model 120-34, using a carbon paste (TU-10s, Asahi Chemical Research Laboratory Co., Ltd., Utsuki-cho, Hachioji, Tokyo) dried at 150 °C for 30 min in air. An average thickness of 5.46 ± 0.51 μm, through a single printing step, was achieved. For the channel layer, ZnO10 and ZnO40 inks were printed through a mesh model 77-55 with a single printing step, achieving an average thickness of 1.39 ± 0.14 μm and 4.60 ± 0.72 μm, respectively. The same mesh was used to print the CSPE layer (average thickness of 18.43 ± 1.23 μm), which was cured at 70 °C for 5 min, followed by UV-exposure for 5 min. Lastly, the silver gate electrode was printed via a 120-34 mesh (average thickness of 1.78 ± 0.60 μm). For that, silver ink (PE-AG-530, from Conductive Compounds, Inc., Hudson, NH, USA) was printed and dried at 70 °C for 5 min, in air.

Prior to printing steps, glass substrates were cleaned in an ultrasonic bath for 5 min, first in acetone and then in isopropanol, after that rinsed off in deionized water (Millipore) and then dried using nitrogen. Regarding the paper substrates, regular printing paper and nanocellulose paper was based on microfibrillated cellulose (referred to as MFC Kraft). MFC Kraft was produced from a Kraft pulp subjected to a mechanical treatment followed by six homogenizing steps at 1500 bar. The nanocellulose paper was prepared by slow casting and evaporation of water from 20 mL of MFC suspension (2 wt % MFC) under ambient conditions in polystyrene Petri dishes (85 mm diameter). The resulting paper substrates had a thickness of approximately 62.4 μm, estimated from the average of five measurements made using a Mitutoyo digital micrometer.

The thickness of the screen-printed layers was measured with an Ambios XP-Plus 200 Stylus profilometer. The channel length (L) and width (W) of the EGTs were measured using a Leica IC80 HD microscope and LAS V4.3 software. The EGTs were electrically analyzed in the dark at room temperature using a microprobe station (M150 Cascade Microtech, Beaverton, OR, USA) connected to a semiconductor parameter analyzer (4155C Agilent, Santa Clara, CA, USA) controlled by the software Metrics ICS.

Nanoindentation experiments were performed using a Hysitron TI 950 Triboindenter tool with a Berkovich indenter on three different sample regions, with the goal to characterize the films mechanically. For each region, the indents were performed with a maximum load, P_{max}, of 1000 μN, 2500 μN, and 5000 μN. For each P_{max}, 20 indents were made, with a 5 s loading time, a 10 s creep time, and finally a 5 s unloading time.

3. Results and Discussion

Figure 1a depicts the process steps of the fully printed EGTs arranged in a staggered-top gate architecture. Figure 1b,d show the topographical SEM surface image of screen-printed layers on glass starting from ZnO inks with different NPs content (for details regarding ink formulation, please see "Materials and Methods" section). The morphology is similar, despite the visible agglomerates for the lowest ZnO content (Figure 1c,e). The RMS surface roughness values obtained by AFM (Figure S1) are 115.1 nm and 132.8 nm for layers printed from ZnO10 (10 wt % ZnO NPs) and ZnO40 (40 wt % ZnO NPs) inks, respectively. It is known that high surface roughness can heavily influence the interface quality, compromising the device performance [15]. Nonetheless, the rheologic characteristics of the ink developed here provided a good packing and self-leveling of the NPs, resulting in excellent film compactness and substrate coverage, ensuring continuous conduction paths, as required for fully functional devices.

Since the target is to fabricate EGTs on paper substrates, the mechanical properties of individual ZnO40 screen-printed films were studied after being submitted to bending cycles. Figure S2 shows that the electrical resistance of the screen-printed ZnO40 films on paper continuously increases after 100, 500, and 1000 bending cycles. Crack formation (see Figure S3) observed after 1000 bending cycles helps to explain this. Nanoindentation was performed in three different regions of the ZnO screen-printed layers subjected to the 1000 bending cycles to determine Young´s modulus (E) and the indentation hardness (H) (see Figure S4). For the analysis, the maximum penetration depth (h_c), was set around 2 μm, i.e., less than 10% of the thickness of the layer, to avoid effects of the paper substrate [16]. Both E and H decrease with increased h_c, which is related with the fluence of the non-sintered ZnO NPs during the indentation. The non-ideal cohesion between them also explains why the E and H are consistently lower after bending cycles.

The ionic and electrochemical response of the CSPE used in this work was determined using electrochemical impedance spectroscopy (EIS). For further details about CSPE characterization please see Figure S5. Figure 2a,b show the experimental data and the fitting obtained from the equivalent circuit model (ECM) for both Bode and Nyquist plots. EDL formation (stabilizing effective capacitance, C_{eff}) occurs for frequencies below 25 Hz, determining the transition from the resistive to the capacitive regime (phase angle equal to −45°). This is normally also a good indicator about the cut-off frequency [17], which is usually a limitation of the EGTs, namely those using solid-state electrolytes.

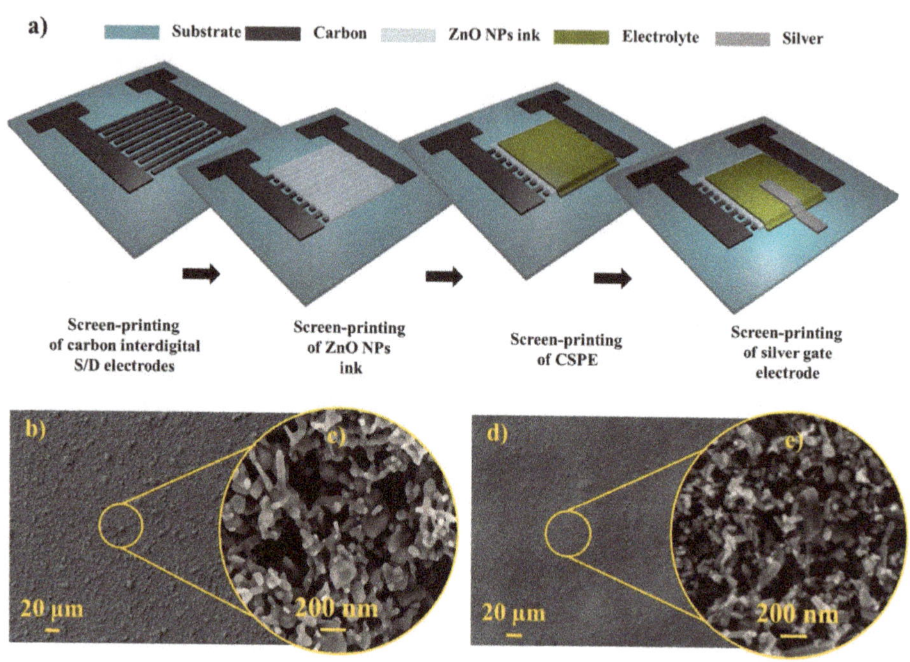

Figure 1. (**a**) Schematic representation of the fabrication steps for the fully printed ZnO electrolyte-gated transistors (EGTs). (**b**,**c**) Topographical view of the screen-printed layer using ZnO10 ink. (**d**,**e**) Topographical view of the screen-printed layer using ZnO40 ink.

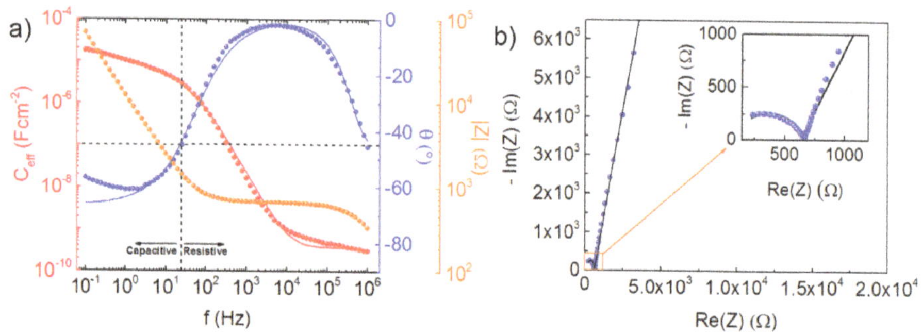

Figure 2. (**a**) The measured effective capacitance, phase, and impedance variation with frequency and (**b**) the respective Nyquist plot of the composite solid polymer electrolyte (CSPE).

C_{eff} gradually stabilizes at a value of about 20 μF cm^{-2} for low frequencies. Some authors use this value as the C_{DL} for further calculations, which is a conservative approach if it is not possible to fit the impedance data with the model, leading to underestimated mobility values, for instance. However, it does not consider contact resistances or the surface inhomogeneities, leading to erroneous values. Here we use Equation (1), proposed by Jovic et al. [18].

$$C_{DL} = \left[Y_0 R_{ext}^{-(\alpha-1)}\right]^{1/\alpha} \tag{1}$$

where Y_0 is linked to the capacitance of the CPE, and α is a constant (between 0 and 1), often referred to as the fractal surface character of the interface, denoting how non-ideally the CPE behaves. The value of

C_{DL} was determined at 2.47 μF cm^{-2}. It is important to note that countless factors are involved, such as the cell setup and dimensions, or even the ECM fitting parameters, which also have an associated error. Nevertheless, the determined values are in the typical range for the use of solid electrolytes in EGTs, i.e., 1–10 μF cm^{-2}, as described by Kim et al. [2].

The transfer characteristics (drain-source current (I_{DS}) vs. gate-source voltage (V_{GS})) of the fully printed ZnO NP EGTs on glass and on paper substrates are shown in Figure 3a,c, respectively. Apart from the ZnO concentration in the ink (10 and 40 wt %) and the substrate used, the device characterization was focused on the influence of the post-printing annealing temperature of the semiconducting inks. The consequent removal of the binder (EC) would ultimately lead to the desired electrical properties, given by the ZnO NPs. As evidenced by TG-DSC (see Figures S6 and S7), temperatures above 350 °C are high enough to fully degrade the binder. Taking this temperature into account, Figure 3a depicts the I_{DS}–V_{GS} transfer characteristic for the ZnO10$_{350\,°C\,Glass}$ EGT. These devices printed on glass substrates show n-type and normally off behavior. Once in the on-state, they reach an I_{DS} value in the range of 10 μA, with an on/off current (I_{on}/I_{off}) ratio of three orders of magnitude, owing to the intrinsic high double layer capacitance of the CSPE [19]. On the other hand, no modulation was observed if the annealing temperature decreases below 350 °C.

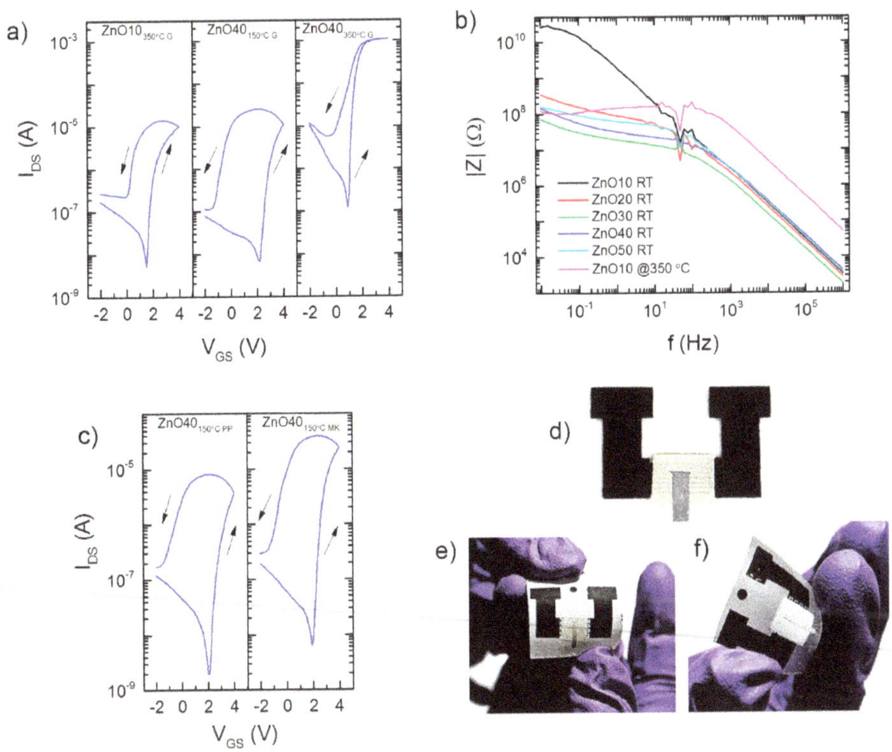

Figure 3. (**a**) Representative transfer characteristics (I_{DS}–V_{GS}) of the ZnO NP EGTs on glass, namely ZnO10 EGT annealed at 350 °C (ZnO10$_{350\,°C\,Glass\,(G)}$); the ZnO40 EGT under processing temperatures of 150 °C (ZnO40$_{150\,°C\,G}$) and the ZnO40 EGT annealed at 350 °C (ZnO40$_{350\,°C\,G}$), respectively. (**b**) Impedance |Z| of the developed ZnO NPs inks, i.e., ZnO10, -20, -30, -40, and -50 measured without annealing (RT), and for the ZnO10 ink after annealing at 350 °C. (**c**) The I_{DS}–V_{GS} of the ZnO40 EGT on printing paper (ZnO40$_{150\,°C\,PP}$) and the ZnO40 EGT on MFC Kraft (ZnO40$_{150\,°C\,MK}$), both at processing temperatures of 150 °C. An image of the printed EGTs on (**d**) glass, (**e**) printing paper, and (**f**) MFC Kraft.

Aiming to reduce the annealing temperature, the ZnO nanoparticle concentration in the ink was increased up to 40 wt %. As depicted on Figure 3a, proper optimization of the NP content allows for electrical modulation at annealing temperatures of only 150 °C ($ZnO40_{150\,°C\,Glass}$). Higher NP content in the ink promotes more conductive paths between the S/D electrodes, without compromising the printability and layer formation. It is worth mentioning here the advantage of using screen-printing since it enables one to print in a wide range of viscosities, allowing the increase in the ZnO NPs content in the ink. Doing so, the resulting printed ZnO layer have lower impedance $|Z|$, as shown in Figure 3b, suitable for use as a channel in EGTs. Thus, by increasing the ZnO NP concentration, the impedance measured for printed patterns using the ZnO40 ink without annealing, is similar to the one measured when using ZnO10 ink after annealing at 350 °C. Consequently, the performance of EGTs based on ZnO40 ink annealed at 150 °C is similar to that obtained for devices using the ZnO10 ink annealed at 350 °C, with an on/off ratio in the same order of magnitude, a saturation mobility (μ_{Sat}) of 0.02 cm^2 $(Vs)^{-1}$, but a higher anti-clockwise hysteresis. The saturation mobility (μ_{Sat}) was determined using Equation (S1), as described in the Supporting Information.

It can be inferred that, without eliminating totally the EC binder, it may behave as an extension of the electrolyte into the semiconducting layer. Previous studies have shown that cellulose can have electrolyte-like behavior in paper transistors [20–24]. Thus, if the EC is not totally eliminated and remains in the printed layer, electrolytic behavior is extended to the ZnO NP channel. The ZnO matrix stays intact as the current modulation is observed. However, ions can penetrate or be trapped in the EC matrix and remain in the vicinity of the semiconductor NPs, even when sweeping back to negative V_{GS}, resulting in ion trapping. The hypothesis of electrolyte extension is sufficiently supported by the results obtained for the EGT based on ZnO40 ink annealed at 350 °C printed on glass ($ZnO40_{350°C\,Glass}$ EGTs). Its I_{DS}–V_{GS} characteristics highlight a considerable reduction in the anti-clockwise hysteresis. This fact is expected since the EC is burned-out, as already explained, so the ion trapping phenomena within the channel layer might be highly reduced. Moreover, a clear increase in both I_{off} and I_{on} of the device is visible from the same I_{DS}–V_{GS} characteristics. This increase is related to an enhanced channel conductivity, resulting from the improved charge transport between the ZnO NPs due to the thermal degradation of EC. Again, the higher NP concentration in the ZnO40 ink in comparison with the ZnO10 one leads to a higher NP density in the printed films. Moreover, partial sintering of the NPs, during the burn-out process, should not be excluded, also in relation to the increasing I_{Off} and I_{On}. In fact, it is known that the sintering temperatures of nanosized structures owing to the high surface-to-volume ratio are considerably lower compared to the bulk material [25]. As a result, the $ZnO40_{350\,°C\,Glass}$ EGT stands out with a μ_{Sat} of 5.73 cm^2 $(Vs)^{-1}$, an I_{On}/I_{Off} ratio of almost four orders of magnitude, and reduced hysteresis, less than 1 V. The performance of these devices compares well with reported values for nanoparticle-based EGTs. Bong et al. [26] reported on ZnO transistors annealed at 280 °C, using an ion-gel as electrolyte (drop-casted) with a μ_{Sat} of 12.1 cm^2 $(Vs)^{-1}$ and an I_{on}/I_{off} ratio of about 10^5. Santos et al. [11] have reported spin-coated gallium-indium-zinc-oxide (GIZO) NP EGTs, with I_{on}/I_{off} ratios between 10^3 and 10^6 and a μ_{Sat} between 6×10^{-3} and 1 cm^2 $(Vs)^{-1}$. Likewise, Dasgupta et al. [15] have reported inkjet printed indium-tin-oxide (ITO) EGTs, with an I_{on}/I_{off} ratio of about 10^4 and a mobility of 5 cm^2 $(Vs)^{-1}$, after annealing at 400 °C.

It is worth mentioning that the μ_{Sat} here obtained for the fully printed ZnO EGTs may be influenced by surface roughness. Consequently, it generates charge traps at the ZnO/electrolyte as well as increases the contact resistance at S/D electrodes, as suggested by Dasgupta et al. [15].

Finally, we could reproduce our ZnO40 EGTs on conventional printing paper ($ZnO40_{150\,°C\,PP}$ EGT, Figure 3c,e) and MFC Kraft substrates ($ZnO40_{150\,°C\,MK}$, Figure 3c,f), since they can withstand temperatures up to 150 °C. The difference in the μ_{Sat}, (0.01 and 0.07 cm^2 $(Vs)^{-1}$) for both paper substrates, depicted in Table 1, can be explained by the influence of the paper substrates roughness. Via 3D profilometry (see Figure S8), the scanned area (500 μm × 500 μm) of the conventional printing paper and MFC Kraft yields an average surface roughness of 2.98 and 1.70 μm, respectively. As verified earlier, the high surface roughness of the substrate owing to the poor dispersion of the ZnO NPs contributes to a large interface roughness between the EGT channel and the CSPE. This result confirms conclusions drawn by Dasgupta et al. [15]. Consequently, the ineffective gating leads to non-uniform electric fields in the semiconducting layer, originating lower μ_{Sat}. Moreover, increased S/D contact resistance can also affect the measured transfer characteristics and consequently the calculated carrier mobility.

Table 1. Summary of the electrical characterization of the produced fully printed ZnO NP EGTs: channel width and length ratio (W/L), the drain voltage (V_{DS}), the turn on voltage (V_{on}), the on-off current ratio (I_{on}/I_{off}), the subthreshold swing (SS), the transconductance (gm), and the saturation mobility (μ_{Sat}).

Device Designation	W/L	V_{DS} (V)	V_{on} (V)	I_{on}/I_{off}	SS (V dec^{-1})	g_m (S)	μ_{Sat} (cm^2 (Vs)$^{-1}$)
ZnO10$_{350\ °C\ G}$ EGT	120	1.2	1.50	2.00×10^3	0.08	2.58×10^{-6}	0.02
ZnO40$_{150\ °C\ G}$ EGT	120	1.1	2.20	1.55×10^3	0.11	3.55×10^{-6}	0.02
ZnO40$_{350\ °C\ G}$ EGT	120	0.9	0.90	8.74×10^3	0.06	8.44×10^{-4}	5.73
ZnO40$_{150\ °C\ PP}$ EGT	120	1.0	2.10	1.73×10^3	0.01	1.41×10^{-6}	0.01
ZnO40$_{150\ °C\ MK}$ EGT	120	1.3	1.90	3.72×10^3	0.21	1.08×10^{-5}	0.07

All the ZnO NP EGTs reported here exhibit extremely low subthreshold slope (SS), showing the rapidness of the ion accumulation process. Additionally, the characteristic low-voltage operation between −2 and 4 V, for this device technology (EGTs), was confirmed. Besides that, the dynamic characterization of the ZnO40$_{150\ °C\ MK}$ EGT (see Figure S9) showed that the device operates between 0.01 and 10 Hz without reaching the cut-off, in accordance with the capacitance/frequency dependence, keeping still above one order of magnitude in the I_{on}/I_{off} ratio.

4. Conclusions

In summary, this study reports a fully printed and flexible inorganic EGT on paper. Besides the low operation voltages, the devices have n-type behavior with an SS of 0.21 V dec^{-1}, a μ_{Sat} of 0.07 cm^2 V^{-1} s^{-1}, and an I_{on}/I_{off} ratio of 3.72×10^3. Investigation in this field could lead to the effective roll-to-roll fabrication of printed logic circuits, with great implications for industrial printing standards, including sustainable and eco-friendly ink formulation. In this context, electrolyte gating, using printable electrolytes, is crucial for the development of printed transistors. In the near future, we foresee printed EGTs to be omnipresent in a wide range of applications, reaching from biosensors, smart packaging to wearable electronics all based on an ultra-low-cost and disposable/recyclable platforms.

Supplementary Materials: The following are available online at http://www.mdpi.com/2079-4991/9/2/169/s1, Figure S1: AFM topographic images (without being annealed) of a) screen-printed ZnO10 ink and b) screen-printed ZnO40, Figure S2: Dependence of the resistance on the number of bending cycles, for screen-printed ZnO40 films on paper, Figure S3: Cross section image of the ZnO40 screen-printed film on paper, after being subjected to 1000 bending cycles. The cross section was prepared by FIB, Figure S4: Young´s Modulus and Hardness results obtained for 3 different regions, for screen-printed ZnO40 films on paper after being subjected to the bending cycles, Figure S5: (a) Schematic representation of the electrochemical cell setup with an area (A) of 1.04 cm^2 and (b) the corresponding ECM suggested by Dasgupta et al. adapted from [15] where R_{ext} is contact resistance, R_b and C_b corresponds to the electrolyte resistance and capacitance, respectively, and CPE the constant phase element, and (c) the AFM topographic image of the screen-printed composite solid polymer electrolyte (CSPE), Figure S6: DSC-TG curves performed in air up to 500 °C for the Ethyl Cellulose (EC) powder, Figure S7: DSC-TG curves performed in air up to 500 °C for the binder solution (ethyl cellulose 5 wt% dissolved in a solvent mixture of toluene:ethanol (80:20 %v/v), Figure S8: The 3D profilometry scan of the surface of the (a) printing paper and (b) MFC kraft substrates, Figure S9: Dynamic electrical characterization of the ZnO40$_{150\ °C\ MK}$ EGT: variation of the I_{On}/I_{Off} using a square-shaped V_G (V_{DS} = 1.3 V, 5 cycles for each frequency 0.01, 0.1, 0.25, 0.5 and 1 Hz).

Author Contributions: J.T.C. performed most of the laboratory work, methodology, data analysis, and writing—original draft preparation. V.D. conducted the bending tests. P.G. and I.C. performed the printing processes and electrolyte preparation. E.F. and R.M. participated in funding acquisition and reviewed and edited the paper. A.C. and E.Z. performed the mechanical characterizations, electron microscopy, and formal analysis, and they reviewed and edited the paper. L.P. is credited for conceptualization, methodology, data analysis, supervision, and funding acquisitions and reviewed and edited the paper.

Funding: This research was funded by FCT—Portuguese Foundation for Science and Technology—through the Ph.D. scholarships SFRH/BD/139225/2018 (José Tiago Carvalho), SFRH/BD/125191/2016 (Paul Grey), and SFRH/BD/126409/2016 (Inês Cunha). The authors would like to acknowledge the European Commission under project NewFun (ERC-StG-2014, GA 640598) and BET-EU (H2020-TWINN-2015, GA 692373). This work was also supported by the FEDER funds through the COMPETE 2020 Program and the National Funds through the FCT under the Project No. POCI-01-0145-FEDER-007688, Reference UID/CTM/50025, project PapEl, reference PTDC/CTM-NAN/5172/2014.

Acknowledgments: The authors would like to acknowledge Martin Gall for the supervision of Viorel Dubceac while at FhG-IKTS, to Daniela Gomes for some of the SEM images and to Ana Pimentel for the DSC-TG analysis.

Conflicts of Interest: The authors declare no conflict of interest.

References

1. Khan, S.; Lorenzelli, L.; Dahiya, R.; Member, S. Technologies for Printing Sensors and Electronics over Large Flexible Substrates: A Review. *IEEE Sens. J.* **2015**, *15*, 3164–3185. [CrossRef]
2. Kim, S.H.; Hong, K.; Xie, W.; Lee, K.H.; Zhang, S.; Lodge, T.P.; Frisbie, C.D. Electrolyte-Gated Transistors for Organic and Printed Electronics. *Adv. Mater.* **2013**, *25*, 1837. [CrossRef]
3. Herlogsson, L. Electrolyte-Gated Organic Thin-Film Transistors. Ph.D Dissertation, Linköping Universitet Institute of Technology, Linköping, Sweden, 2011.
4. Tate, J.; Rogers, J.A.; Jones, C.D.W.; Vyas, B.; Murphy, D.W.; Li, W.; Bao, Z.; Slusher, R.E.; Dodabalapur, A.; Katz, H.E. Anodization and Microcontact Printing on Electroless Silver: Solution-Based Fabrication Procedures for Low-Voltage Electronic Systems with Organic Active Components. *Langmuir* **2000**, *16*, 6054–6060. [CrossRef]
5. Inoue, Y. Organic thin-film transistors based on anthracene oligomers. *J. Appl. Phys.* **2004**, *95*, 5795–5799. [CrossRef]
6. Yu, X.J.; Xu, J.B.; Cheung, W.Y.; Ke, N. Optimizing the growth of vanadyl-phthalocyanine thin films for high-mobility organic thin-film transistors. *J. Appl. Phys.* **2007**, *102*. [CrossRef]
7. Hong, K.; Kim, S.H.; Lee, K.H.; Frisbie, C.D. Printed, sub-2V ZnO Electrolyte Gated Transistors and Inverters on Plastic. *Adv. Mater.* **2013**, *25*, 3413–3418. [CrossRef] [PubMed]
8. Park, S.; Lee, S.; Kim, C.-H.; Lee, I.; Lee, W.-J.; Kim, S.; Lee, B.-G.; Jang, J.-H.; Yoon, M.-H. Sub-0.5 V Highly Stable Aqueous Salt Gated Metal Oxide Electronics. *Sci. Rep.* **2015**, *5*, 13088. [CrossRef] [PubMed]
9. Bandiello, E.; Sessolo, M.; Bolink, H.J. Aqueous electrolyte-gated ZnO transistors for environmental and biological sensing. *J. Mater. Chem. C* **2014**, *2*, 10277–10281. [CrossRef]
10. Wang, D.; Noël, V.; Piro, B. Electrolytic Gated Organic Field-Effect Transistors for Application in Biosensors—A Review. *Electronics* **2016**, *5*, 9. [CrossRef]
11. Santos, L.; Nunes, D.; Calmeiro, T.; Branquinho, R.; Salgueiro, D.; Barquinha, P.; Pereira, L.; Martins, R.; Fortunato, E. Solvothermal Synthesis of Gallium–Indium-Zinc-Oxide Nanoparticles for Electrolyte-Gated Transistors. *ACS Appl. Mater. Interfaces* **2015**, *7*, 638–646. [CrossRef]
12. Barquinha, P.; Pereira, S.; Pereira, L.; Wojcik, P.; Grey, P.; Martins, R.; Fortunato, E. Flexible and Transparent WO_3 Transistor with Electrical and Optical Modulation. *Adv. Electron. Mater.* **2015**, *1*, 1500030. [CrossRef]
13. Rose, A. Considerations in Formulation and Manufacturing of Thick Film Inks. *Electrocompon. Sci. Technol.* **1981**, *9*, 43–49. [CrossRef]
14. Inukai, K.; Takahashi, Y.; Ri, K.; Shin, W. Rheological analysis of ceramic pastes with ethyl cellulose for screen-printing. *Ceram. Int.* **2015**, *41*, 5959–5966. [CrossRef]
15. Dasgupta, S.; Stoesser, G.; Schweikert, N.; Hahn, R.; Dehm, S.; Kruk, R.; Hahn, H. Printed and Electrochemically Gated, High-Mobility, Inorganic Oxide Nanoparticle FETs and Their Suitability for High-Frequency Applications. *Adv. Funct. Mater.* **2012**, *22*, 4909–4919. [CrossRef]
16. Antunes, J.M.; Fernandes, J.V.; Sakharova, N.; Oliveira, M.C.; Menezes, L.F. On the determination of the Young's modulus of thin films using indentation tests. *Int. J. Solids Struct.* **2007**, *44*, 8313–8334. [CrossRef]
17. Yuan, H.; Shimotani, H.; Ye, J.; Yoon, S.; Aliah, H.; Tsukazaki, A.; Kawasaki, M.; Iwasa, Y. Electrostatic and Electrochemical Nature of Liquid-Gated Electric-Double-Layer Transistors Based on Oxide Semiconductors. *J. Am. Chem. Soc.* **2010**, *132*, 18402–18407. [CrossRef]

18. Jović, V.D.; Jović, B.M. EIS and differential capacitance measurements onto single crystal faces in different solutions: Part I: Ag(111) in 0.01 M NaCl. *J. Electroanal. Chem.* **2003**, *541*, 1–11. [CrossRef]
19. Nasr, B.; Wang, D.; Kruk, R.; Rösner, H.; Hahn, H.; Dasgupta, S. High-Speed, Low-Voltage, and Environmentally Stable Operation of Electrochemically Gated Zinc Oxide Nanowire Field-Effect Transistors. *Adv. Funct. Mater.* **2013**, *23*, 1750–1758. [CrossRef]
20. Fortunato, E.; Correia, N.; Barquinha, P.; Pereira, L.; Goncalves, G.; Martins, R. High-Performance Flexible Hybrid Field-Effect Transistors Based on Cellulose Fiber Paper. *IEEE Electron Device Lett.* **2008**, *29*, 988–990. [CrossRef]
21. Pereira, L.; Gaspar, D.; Guerin, D.; Delattre, A.; Fortunato, E.; Martins, R. The influence of fibril composition and dimension on the performance of paper gated oxide transistors. *Nanotechnology* **2014**, *25*, 094007. [CrossRef]
22. Gaspar, D.; Fernandes, S.N.; de Oliveira, A.G.; Fernandes, J.G.; Grey, P.; Pontes, R.V.; Pereira, L.; Martins, R.; Godinho, M.H.; Fortunato, E. Nanocrystalline cellulose applied simultaneously as the gate dielectric and the substrate in flexible field effect transistors. *Nanotechnology* **2014**, *25*, 094008. [CrossRef] [PubMed]
23. Grey, P.; Gaspar, D.; Cunha, I.; Barras, R.; Carvalho, J.T.; Ribas, J.R.; Fortunato, E.; Martins, R.; Pereira, L. Handwritten Oxide Electronics on Paper. *Adv. Mater. Technol.* **2017**, *2*, 1700009. [CrossRef]
24. Cunha, I.; Barras, R.; Grey, P.; Gaspar, D.; Fortunato, E.; Martins, R.; Pereira, L. Reusable Cellulose-Based Hydrogel Sticker Film Applied as Gate Dielectric in Paper Electrolyte-Gated Transistors. *Adv. Funct. Mater.* **2017**, *27*, 1606755. [CrossRef]
25. Kamyshny, A.; Magdassi, S. Conductive Nanomaterials for Printed Electronics. *Small* **2014**, *10*, 3515–3535. [CrossRef]
26. Bong, H.; Lee, W.H.; Lee, D.Y.; Kim, B.J.; Cho, J.H.; Cho, K. High-mobility low-temperature ZnO transistors with low-voltage operation. *Appl. Phys. Lett.* **2010**, *96*, 2008–2011. [CrossRef]

nanomaterials

MDPI

Article

Multi-Level Cell Properties of a Bilayer Cu_2O/Al_2O_3 Resistive Switching Device

Jonas Deuermeier [1], Asal Kiazadeh [1,*], Andreas Klein [2], Rodrigo Martins [1] and Elvira Fortunato [1]

1 i3N/CENIMAT, Department of Materials Science, Faculty of Science and Technology, Universidade NOVA de Lisboa and CEMOP/UNINOVA, Campus de Caparica, 2829-516 Caparica, Portugal; j.deuermeier@campus.fct.unl.pt (J.D.); rfpm@fct.unl.pt (R.M.); emf@fct.unl.pt (E.F.)

2 Department of Materials and Earth Sciences, Technische Universität Darmstadt, Otto-Berndt-Straße 3, D-64287 Darmstadt, Germany; aklein@esm.tu-darmstadt.de

* Correspondence: a.kiazadeh@fct.unl.pt; Tel.: +351212948562

Received: 31 December 2018; Accepted: 14 February 2019; Published: 19 February 2019

Abstract: Multi-level resistive switching characteristics of a Cu_2O/Al_2O_3 bilayer device are presented. An oxidation state gradient in copper oxide induced by the fabrication process was found to play a dominant role in defining the multiple resistance states. The highly conductive grain boundaries of the copper oxide—an unusual property for an oxide semiconductor—are discussed for the first time regarding their role in the resistive switching mechanism.

Keywords: resistive switching memories; multi-level cell; copper oxide; grain boundaries; aluminum oxide

1. Introduction

Two terminal resistive switches have motivated many studies on nanoscale data storage and neuromorphic applications due to their superior performance including high density, simple structure, fast programing, long retention, and low power consumption [1]. Conductive bridging random access memories (CBRAMs) are resistive switches which contain an electrochemically active metal electrode (Ag, Cu). It is widely accepted that programable metallization on the nanoscale within the switching medium causes bistable properties. To better control the localization and the diameter of conducting filaments (CFs) and improve the resistive switching uniformity and stability, bilayer devices (based on chalcogenides or oxides) consisting of a switching layer and a buffer layer were proposed some years ago [2,3]. Note that oxide bilayers are well compatible with complementary metal-oxide-semiconductor (CMOS) technology.

Among numerous oxide-based switching materials, Cu_2O and Al_2O_3 have advantageous properties such as being abundant, of low cost, having low processing temperatures, are environmentally friendly, and in the case of Cu_2O, electrochemically active. Aluminum oxide is widely applied as a dielectric of thin film transistors [4,5], thus a resistive switching device based on Cu_2O/Al_2O_3 may decrease the processing steps for circuits such as one transistor-one resistor cells (1T-1R).

Various examples of resistive switching with a single Cu_2O layer sandwiched between two electrodes have been reported. Starting in 1969, the disproportionation of Cu_2O into Cu and CuO was considered to give rise to a copper filament [6,7]. In 1982 it was discussed that the Cu/Cu_2O Schottky barrier was beneficial in creating reproducible switching events since it generates a region of comparatively high electric field [8]. However, the current blocking effect of a high barrier can also prevent switching from happening [9]. The most recent discussion of the switching mechanism concludes a filament of copper vacancies in the high conductance state and a Cu/Cu_2O Schottky barrier in the low conductance state [10]. Others have reported multiple filaments of metallic copper;

however, using devices with higher cell size and higher current than in the aforementioned work [11]. Note, that polycrystalline Cu_2O was found to show highly conductive grain boundaries due to the presence of nanocrystalline CuO [12], causing the macroscopic conductivity to be corelated with the grain boundary density [13]. Another material in which defect segregation and increased conductivity in the grain boundary play a dominant role in resistive switching is HfO_2 [14,15].

Thick insulators like Al_2O_3 and SiO_2 do not conduct ions, but behave as electrolytes when sufficiently thin [16,17]. The result is a metallic filament growth from the inert electrode towards the active electrode. An opposite growth direction has also been reported both for Al_2O_3 [18] and SiO_2 [19,20]. This mechanism appears to be favored when the metal is embedded as nanoclusters within the switching matrix.

Two distinct types of low resistance states (LRS) were observed in a ZrO_2/Cu-based memory, which were accessible by different set voltages. At low set voltages, an ionic filament based on defects in the switching matrix ZrO_2 is activated, whereas at a higher set voltage a metallic filament based on copper is formed. The ionic filament is reset in a unipolar fashion, whereas the metallic one is bipolar [21]. In another bilayer memory device based on AlO_x/WO_x, multi-level cell (MLC) programing was accessed by controlling the current compliance (CC) during the set operation. The MLC approach decreased the processing cost of bits/cells. The conduction mechanisms in the LRS were found to be metallic for high CC, electron-hopping between metallic precipitates for intermediate CC and Schottky emission at the Al/AlO_x interface for low CC and low temperature [22]. Another approach to control the metallic filament formation in a CBRAM device is to limit the supply of the reactive species by insertion of a buffer layer or by alloying the copper with another element [23–25].

This work presents the resistive switching behavior of Al_2O_3/Cu_2O bilayer devices. Taking into account previously published knowledge on the structural and electrical inhomogeneity of polycrystalline Cu_2O [12,13], as well as on the interface between Al_2O_3 and Cu_2O [26], allows to clarify the roles of the individual layers in the resistive switching observed in the devices.

2. Materials and Methods

The device structure is schematically represented in Figure 1. Polycrystalline Cu_2O (93 nm) was deposited by reactive magnetron sputtering from a metallic copper target (Kurt J. Lesker Company, Jefferson Hills, PA, USA) on commercial indium-tin oxide (ITO) on glass (Corning Inc., Corning, NY, USA). Subsequently, Al_2O_3 (16 nm) was deposited by atomic layer deposition (ALD) using trimethylaluminum (SAFC Hitech Ltd., Bromborough, UK) and water as precursors. Metallic copper was located between the copper oxide and the aluminum oxide layer as a consequence of the ALD process. Detailed information on the preparation method of these layers can be found in Reference [26]. As top contacts, sputter-deposited platinum (Kurt J. Lesker Company, Jefferson Hills, PA, USA) was used with a diameter of 100 μm, patterned by a shadow mask, using a commercial sputter-coater (Quorum Technologies Ltd., Lewes, UK). The current–voltage (I–V) characteristics of the ITO/Cu_2O/Al_2O_3/Pt devices were measured at room temperature in air using a Keithley 4200-SCS semiconductor parameter analyzer (Keithley Instruments LLC, Cleveland, OH, USA) connected to a Janis ST-500 probe station (Janis Research Company LLC, Woburn, MA, USA). The bias was applied to the top electrode. The delay time until forming was measured under application of a constant voltage in sampling mode with a sampling interval of 0.5 s, using a measurement speed optimized for low noise and high accuracy (delay factor 1.3, filter factor 3, automatic A/D integration time setting). Temperature-dependent measurements were done in vacuum using liquid nitrogen for cooling, controlled by a Lake Shore 336 temperature controller (Lake Shore Cryotronics Inc., Westerville, OH, USA).

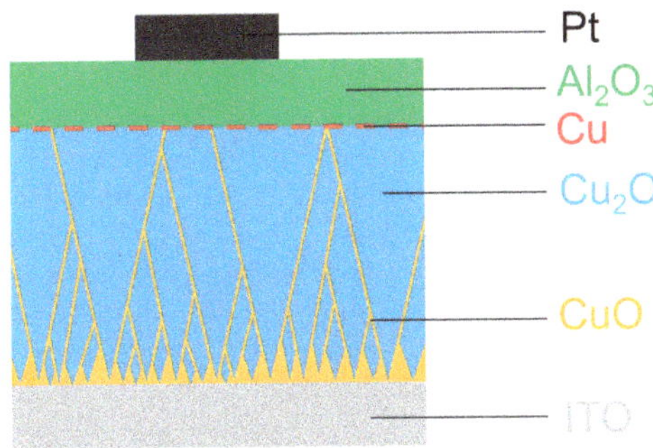

Figure 1. Device structure in cross-sectional view. The inhomogeneities in the copper oxide are schematically shown in orange (CuO-containing grain boundaries) and red (metallic copper at the interface between Al_2O_3 and Cu_2O).

3. Results

At first, the resistive switching template was tentatively formed by an electroforming process with low CC of 500 nA. Such a low current compliance allows for an extremely low power consumption of memory cells. The negative bias in voltage sweep mode was required to form the resistive switching memory device. The typical I–V behavior of the electroforming process is shown in Figure 2a. The memory cell reverted to the high resistance state (HRS) only by applying a positive voltage (reset) and the forming voltage was always higher than the required voltage for the set operation. The set and reset operating voltages are very low, <±1.5 V. An LRS/HRS ratio of 10^2 was obtained at the first programming cycles, which was reduced and stabilized to ~10 after continued cycling. Initially, the HRS was as low as the pristine state (i.e., the electrical characteristic of the as-fabricated device), mainly due to the full rupture of initial narrow filament or a non-complete forming process. With further cycling, the I–V characteristics showed an enhancement of current rectification in the LRS, whereas the HRS current level increased after a few cycles, giving rise to an LRS/HRS margin below one order of magnitude. The self-forming during the initial cycles led to a selector-less device performance, which was highly stable, as represented by the distribution of the resistance states in Figure 2b and the retention data after the stabilization in Figure 2c.

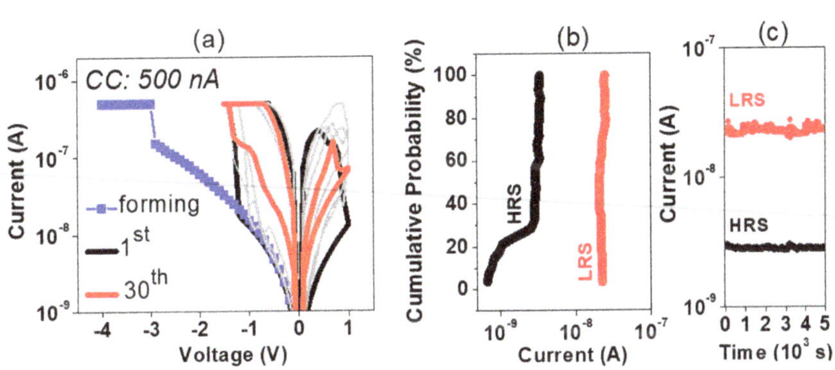

Figure 2. I–V sweeps of forming with a current compliance (CC) of 500 nA (blue squares) and 1st (black) and 30th (red) cycles, as well as a few intermediate cycles in grey (**a**), corresponding distribution of resistance states (**b**) and retention over 5×10^3 s (**c**).

In order to shed light on the pre-forming process (prior to the stabilization during cycling), the delay time t_d until pre-forming under constant voltage was measured for different voltages [27]. Each device was initially in the pristine state and a CC of 500 nA was used (see Figure 3a). Larger delay times were observed for smaller voltages. The delay time had an exponential dependence on voltage (see Figure 3b). A good fitting was obtained with $t_0 = 2.14 \times 10^5$ s and $\gamma = 3.91$. The factor γ is referred to as a voltage acceleration parameter to a soft breakdown process [27] and higher values have already been reported for thinner films [28]. Similar results were reported for the breakage of a thin Al_2O_3 film [27]. This means, that the aluminum oxide plays a dominant role in the pre-forming process.

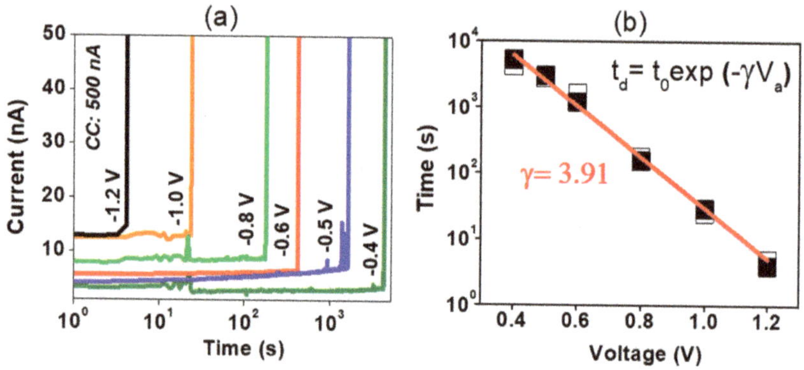

Figure 3. Current with respect to time under different constant voltage bias (**a**). Delay time to pre-forming with respect to voltage (**b**). Two individual measurements are shown by the filled and empty symbols. The exponential fit is shown in red.

A multi-level cell (MLC) operation can be obtained by changing the set and reset conditions. To modulate stable MLC states, a higher LRS/HRS ratio is required to obtain distinguishable conductance states for a long retention time. One way to increase the LRS/HRS ratio is to increase CC. Figure 4a–c show the typical I–V characteristics of different resistance states under application of CCs of 10 μA, 50 μA, and 500 μA, respectively. The distribution of different resistance states during sweeping cycles are shown in Figure 4d. The highest cycling stability was observed for a CC of 50 μA. All resistance states show no degradation in data retention over 5×10^3 s, see Figure 4e. Note, that once the device sensed the 500 μA, the resistance states could not be well tuned for lower CC. The HRS was not dependent on the applied CC, suggesting an identical charge transport mechanism independent on the size of the filament.

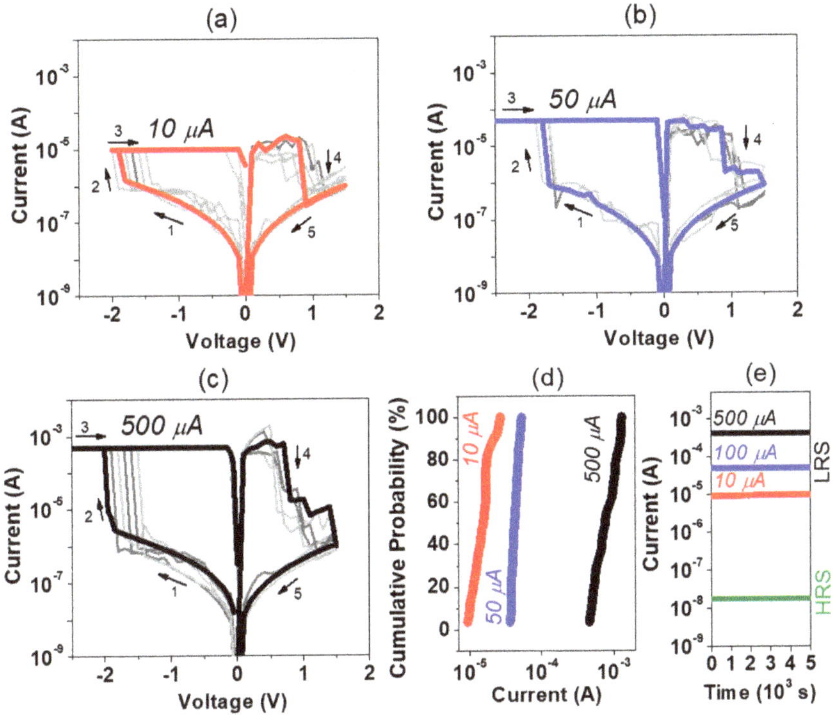

Figure 4. Typical I–V cycles of devices formed at CCs of 10 μA (**a**), 50 μA (**b**), and 500 μA (**c**), respectively, corresponding distribution of low resistance states (LRS) states (**d**) and retention over 5 × 10^3 s (**e**).

Once the device is at a certain LRS, MLC can be also achieved by controlling the reset voltage. Figure 5a shows a typical I–V sweep under modulation of the reset voltage. Different conductance states are achieved after reset stop voltages of 0.5 V, 1 V, and 1.5 V. The conductance state distribution and the data retention of the MLC by controlling the reset voltages is shown in Figure 5b,c, respectively. Besides a good retention for all states, a higher spread of the conductance states was found for the HRS 1, compared to the LRS and HRS 2, which is discussed further below.

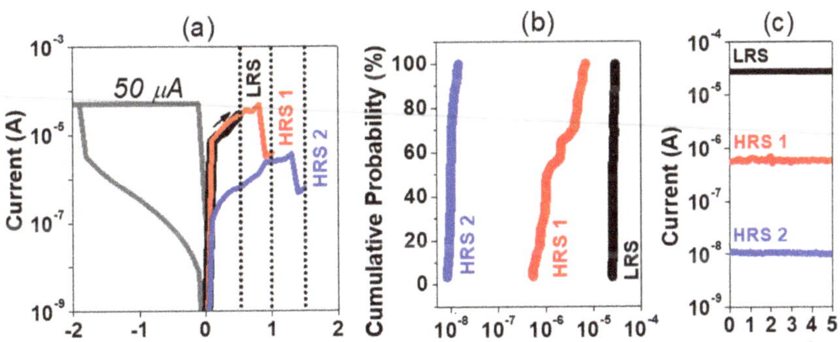

Figure 5. I–V sweeps of set at CC of 50 μA and reset to two distinct high resistance states (HRS), (**a**) with the corresponding distribution of resistance states (**b**) and retention over 5 × 10^3 s (**c**).

To investigate the MLC resistive switching behavior, the temperature dependence of the different states was studied, presented in Figure 6. Arrhenius plots of the HRS and the different LRS at CC 10 μA and 50 μA are shown in Figure 6a. A semiconducting behavior with good linear fits of the Arrhenius plots was obtained for the HRS and the LRS with a CC of 10 μA. The activation energies are E_a = 0.28 eV and E_a = 0.18 eV, respectively. When the CC was set for 50 μA, the current-temperature dependence still shows the semiconducting behavior, but no satisfactory linear fit was obtained, and the activation energy was too low to fulfill the conditions for the Boltzmann approximation. At high CC of 500 μA shown in Figure 6b, the conductance is decreased with the increase of temperature, which confirms a metallic conduction type. A positive temperature coefficient of 1.3×10^{-3} K^{-1} was obtained from fitting the results.

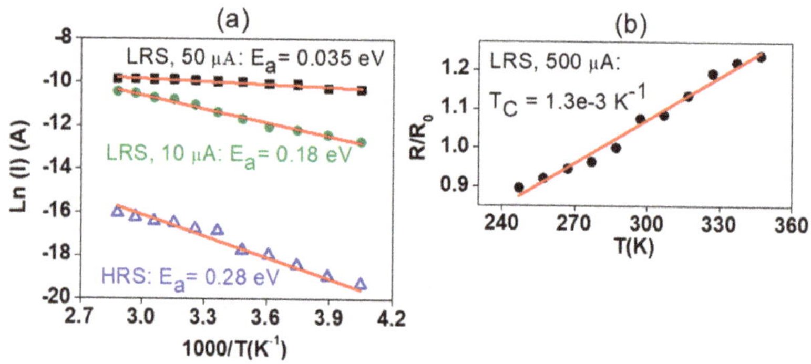

Figure 6. Arrhenius plots with corresponding fits (red lines) of the HRS (blue triangles) and the different LRS at CC of 10 μA (green circles) and 50 μA (black squares) (**a**), linear fit with respect to temperature (red line) of the LRS at CC of 500 μA (black circles) (**b**).

4. Discussion

The mechanism of the conductive filament formation shall now be discussed based on the above presented results. The initial stages of forming are dominated by the soft breakdown of Al_2O_3. Since this was observed with a CC of only 500 nA, the conclusion of an initial forming in the Al_2O_3 holds also for higher CC. Note, that since a similar acceleration parameter has been reported for Al_2O_3 without the presence of a metallic species [27], we cannot be entirely certain about the nature of the filament in Al_2O_3. Due to the presence of metallic copper at the interface between Al_2O_3 and Cu_2O (either as nanoclusters or as very thin film) [26], a filament formation from the interface with Cu_2O towards the Pt electrode is likely [19,20], although the presented electrical data does not allow to be conclusive here. Independently of the filament growth direction, a counter reaction at the Pt electrode is required to fulfill the charge neutrality condition. Since electroforming was conducted in ambient air, moisture is the most probable reactant [29].

With the breakage of aluminum oxide, the resistance values of the MLC LRS were modulated by the filament formation in copper oxide. This conclusion is based on the activation energies given in Figure 6a, which closely resemble typical values of copper oxide (see detailed discussion further below). Metallic copper is not present close to the bottom electrode, so ions are oxidized, which are bound in the copper oxide lattice. Since water is typically reduced at the cathode [29], it is not considered for the anode reaction. Since the copper ions in CuO are fully oxidized (assuming stoichiometric CuO), the oxidation reaction must involve monovalent ions from the Cu_2O grains. The corresponding anode reaction is written in Kroeger–Vink notation:

$$Cu_{Cu}^{\times} \rightarrow Cu_i^{\bullet\bullet} + V_{Cu}' + e'$$

Once the interstitial copper is formed, it becomes a mobile species and migrates towards the cathode, where it is deposited as metallic copper. Copper ions in copper oxide are highly mobile through a vacancy-assisted mechanism [30]. Besides the divalent copper ions, copper vacancies are created. These are the main acceptor defects in Cu_2O [31]. Due to the formation of another phase (metallic copper), the cathode reaction is written as an electrochemical half-reaction.

$$Cu^{2+} + 2e^{-} \rightarrow Cu$$

The growth of the filament during forming in length and probably also in thickness depends on the CC [11]. This is schematically illustrated in Figure 7. Note that the grain boundaries of Cu_2O are significantly more conductive than the grains due to the presence of CuO [12,13]. For this reason, most likely the filament formation involves the grain boundaries, which is considered in the following.

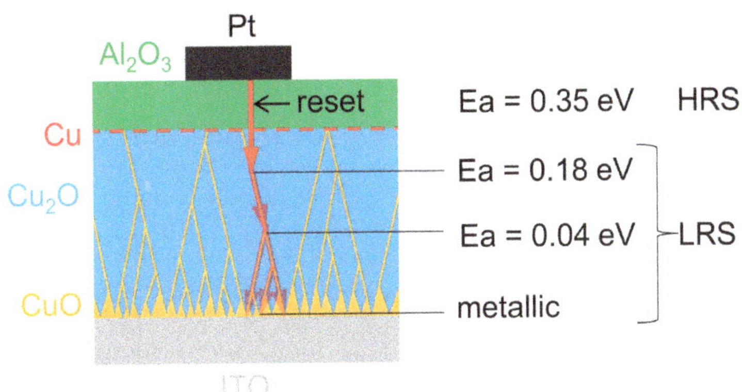

Figure 7. Schematic representation of the switching mechanism at different CCs. The varying extent of the filament inside the copper oxide with different CCs is illustrated by the red arrows.

At lowest CC (500 nA), a Schottky barrier was observed in the positive polarity and the HRS was not stable during the initial cycles. The Schottky barrier clearly shows that copper is in electrical contact with Cu_2O [32]. The instability shows that there are multiple competing highly conductive paths, which most likely correspond to the grain boundaries of Cu_2O. Due to the instability during cycling, the forming of the memory is considered incomplete with CC = 500 nA. When the CC is increased to 10 μA, semiconducting filaments are observed. The activation energy is too low to be related to Cu_2O [33]. A Schottky barrier is not observed anymore. This means that the copper filament is in direct contact with the copper oxide grain boundaries, since the grain boundary has electrical properties similar to CuO [13]. The absence of band bending in CuO causes the non-rectifying contact [34]. Compared to CC = 500 nA, an increased LRS stability is observed which is further increased with an even higher CC of 50 μA. With CC = 50 μA, the activation energy drops below 3kT (which the Boltzmann constant k and the temperature T), most likely due to an increased doping level of the CuO in the grain boundary of Cu_2O and/or the progressed filament growth towards a region of higher grain boundary density close to the substrate interface (see Figure 7). Using a CC of 500 μA, the filament becomes metallic. The temperature coefficient of 1.3×10^{-3} K^{-1} is similar to the one observed in copper nanowires [35]. Hence, this LRS corresponds to a situation in which the metallic copper filament has reached the ITO bottom electrode during forming.

Concerning the reset process, it is important to note that Al_2O_3 has a lower thermal conductivity than Cu_2O [36,37]. This means that the energy created due to Joule heating of the filament dissipated more efficiently in Cu_2O than in Al_2O_3. This leads to a preferential rupture in the Al_2O_3, independent on the CC used during forming. The diffusion of copper is higher in Cu_2O than in Al_2O_3 [38]. Hence, in case the filament in Al_2O_3 is based on metallic copper, it will be thinner in Al_2O_3 compared to Cu_2O.

The comparatively high activation energy in the HRS may be caused by the Al_2O_3 or potentially by a contribution of the Cu_2O near-surface region, which is a region of comparatively high activation energy. Furthermore, as copper ions are released from the filament during reset, they can annihilate copper vacancies in the surrounding Cu_2O grains, causing a decreased intrinsic doping level, thus an increase in activation energy.

With the reset occurring in or close to the Al_2O_3, the set process will also happen here. This means that the switching itself is of pure electrochemical metallization (ECM) type; however, with the difference of having mostly copper oxide as anode material and just a small amount of metallic copper (at the Cu_2O/Al_2O_3 interface). This amount of metallic copper is modulated by the CC during forming, as described earlier. Hence, the copper supply at the Cu_2O/Al_2O_3 interface can impact the switching characteristics similar to devices using Cu_xTe_{1-x} as active electrode [24], which is reflected here by the presented results: low copper amounts (equivalent to low CC) compromise the filament stability (see Figure 2), whereas higher copper amounts (higher CC) require an increased reset voltage (see Figure 4a–c).

5. Conclusions

The resistive switching mechanism of a bilayer system of Cu_2O and Al_2O_3 was investigated. The main objective of this work was to discriminate between the contributions of each individual layer and to discuss how they affect each other. The outcome features valuable indications for future bilayer device design. The observed multi-level operation was controlled by the current compliance. A transition from a semiconducting filament to a metallic filament with increasing current compliance was observed. For the filament growth in Cu_2O, a dual mechanism is proposed, which involves a valence change in the copper oxide at the anode and a metallization reaction at the cathode. The likely preferential filament rupture during reset in the Al_2O_3 layer confines the switching event to Al_2O_3. Consequently, the supply of the active metal is conditioned by the CC-dependent filament growth in the copper oxide during forming, which directly impacts cycling stability.

Author Contributions: Conceptualization, J.D. and A.K. (Asal Kiazadeh); methodology, A.K. (Asal Kiazadeh); validation, J.D. and A.K. (Asal Kiazadeh); formal analysis, A.K. (Asal Kiazadeh); investigation, J.D. and A.K. (Asal Kiazadeh); resources, A.K. (Andreas Klein), E.F. and R.M.; data curation, J.D. and A.K. (Asal Kiazadeh); writing—original draft preparation, J.D.; writing—review and editing, A.K. (Asal Kiazadeh), J.D. and E.F.; visualization, J.D and A.K. (Asal Kiazadeh); supervision, A.K. (Asal Kiazadeh) and E.F.; project administration, R.M. and E.F.; funding acquisition, A.K. (Andreas Klein), E.F. and R.M.

Funding: This research was funded by FEDER funds through the COMPETE 2020 Programme and National Funds through FCT—Portuguese Foundation for Science and Technology under project number POCI-01-0145-FEDER-007688, Reference UID/CTM/50025. J.D. acknowledges funding received from the European Union's Horizon 2020 Research and Innovation Programme through the project HERACLES (Project No. 700395) and the German Science Foundation through the collaborative research center SFB 595 (Electrical Fatigue of Functional Materials). A. Kiazadeh acknowledges FCT for the postdoctoral grant SFRH/BPD/99136/2013 and for funding received through the project NeurOxide (PTDC/NAN-MAT/30812/2017).

Conflicts of Interest: The authors declare no conflict of interest.

References

1. Yang, J.J.; Pickett, M.D.; Li, X.; Ohlberg, D.A.A.; Stewart, D.R.; Williams, R.S. Memristive switching mechanism for metal/oxide/metal nanodevices. *Nat. Nanotechnol.* **2008**, *3*, 429–433. [CrossRef] [PubMed]
2. Jana, D.; Roy, S.; Panja, R.; Dutta, M.; Rahaman, S.Z.; Mahapatra, R.; Maikap, S. Conductive-bridging random access memory: Challenges and opportunity for 3D architecture. *Nanoscale Res. Lett.* **2015**, *10*, 188. [CrossRef] [PubMed]
3. Pei, Y.; Mai, B.; Zhang, X.; Hu, R.; Li, Y.; Chen, Z.; Fan, B.; Liang, J.; Wang, G. Performance improvement of amorphous indium-gallium-zinc oxide ReRAM with SiO_2 inserting layer. *Curr. Appl. Phys.* **2015**, *15*, 441–445. [CrossRef]
4. Robertson, J. High dielectric constant oxides. *Eur. Phys. J. Appl. Phys.* **2004**, *28*, 265–291. [CrossRef]

5. Carlos, E.; Branquinho, R.; Kiazadeh, A.; Barquinha, P.; Martins, R.; Fortunato, E. UV-Mediated Photochemical Treatment for Low-Temperature Oxide-Based Thin-Film Transistors. *ACS Appl. Mater. Interfaces* **2016**, *8*, 31100–31108. [CrossRef] [PubMed]
6. Cook, E.L. Model for the resistive-conductive transition in reversible resistance-switching solids. *J. Appl. Phys.* **1970**, *41*, 551–554. [CrossRef]
7. Zarabi, M.J.; Satyam, M. Switching in copper oxide. *J. Appl. Phys.* **1974**, *45*, 775–780. [CrossRef]
8. Wang, G.J.; Weichman, F.L. The temperature dependence of the electrical conductivity and switching phenomena in Cu_2O single crystals. *Can. J. Phys.* **1982**, *60*, 1648–1655. [CrossRef]
9. Yang, W.Y.; Rhee, S.W. Effect of electrode material on the resistance switching of Cu_2O film. *Appl. Phys. Lett.* **2007**, *91*, 232907. [CrossRef]
10. Yan, P.; Li, Y.; Hui, Y.J.; Zhong, S.J.; Zhou, Y.X.; Xu, L.; Liu, N.; Qian, H.; Sun, H.J.; Miao, X.S. Conducting mechanisms of forming-free $TiW/Cu_2O/Cu$ memristive devices. *Appl. Phys. Lett.* **2015**, *107*, 083501. [CrossRef]
11. Singh, B.; Mehta, B.R.; Varandani, D.; Savu, A.V.; Brugger, J. CAFM investigations of filamentary conduction in Cu_2O ReRAM devices fabricated using stencil lithography technique. *Nanotechnology* **2012**, *23*, 495707. [CrossRef] [PubMed]
12. Deuermeier, J.; Liu, H.; Rapenne, L.; Calmeiro, T.; Renou, G.; Martins, R.; Muñoz-Rojas, D.; Fortunato, E. Visualization of nanocrystalline CuO in the grain boundaries of Cu_2O thin films and effect on band bending and film resistivity. *APL Mater.* **2018**, *6*, 096103. [CrossRef]
13. Deuermeier, J.; Wardenga, H.F.; Morasch, J.; Siol, S.; Nandy, S.; Calmeiro, T.; Klein, A.; Fortunato, E. Highly conductive grain boundaries in copper oxide thin films. *J. Appl. Phys.* **2016**, *119*, 235303. [CrossRef]
14. Bersuker, G.; Yum, J.; Iglesias, V.; Porti, M.; Nafría, M.; Mckenna, K.; Shluger, A.; Kirsch, P.; Jammy, R. Grain boundary-driven leakage path formation in HfO_2 dielectrics. In Proceedings of the 40th European Solid State Device Research Conference, Sevilla, Spain, 14–16 September 2010; pp. 333–336.
15. Lanza, M. A review on resistive switching in high-k dielectrics: A nanoscale point of view using conductive atomic force microscope. *Materials* **2014**, *7*, 2155–2182. [CrossRef] [PubMed]
16. Valov, I.; Waser, R.; Jameson, J.R.; Kozicki, M.N. Electrochemical metallization memories—fundamentals, applications, prospects. *Nanotechnology* **2011**, *22*, 254003. [CrossRef] [PubMed]
17. Hubbard, W.A.; Kerelsky, A.; Jasmin, G.; White, E.R.; Lodico, J.; Mecklenburg, M.; Regan, B.C. Nanofilament Formation and Regeneration during Cu/Al_2O_3 Resistive Memory Switching. *Nano Lett.* **2015**, *15*, 3983–3987. [CrossRef] [PubMed]
18. Celano, U.; Goux, L.; Belmonte, A.; Opsomer, K.; Franquet, A.; Schulze, A.; Detavernier, C.; Richard, O.; Bender, H.; Jurczak, M.; et al. Three-dimensional observation of the conductive filament in nanoscaled resistive memory devices. *Nano Lett.* **2014**, *14*, 2401–2406. [CrossRef]
19. Yang, Y.; Gao, P.; Li, L.; Pan, X.; Tappertzhofen, S.; Choi, S.; Waser, R.; Valov, I.; Lu, W.D. Electrochemical dynamics of nanoscale metallic inclusions in dielectrics. *Nat. Commun.* **2014**, *5*, 1–9. [CrossRef]
20. Tappertzhofen, S.; Hofmann, S. Embedded nanoparticle dynamics and their influence on switching behaviour of resistive memory devices. *Nanoscale* **2017**, *9*, 17494–17504. [CrossRef]
21. Liu, M.; Abid, Z.; Wang, W.; He, X.; Liu, Q.; Guan, W. Multilevel resistive switching with ionic and metallic filaments. *Appl. Phys. Lett.* **2009**, *94*, 233106. [CrossRef]
22. Zhang, Y.; Wu, H.; Bai, Y.; Chen, A.; Yu, Z.; Zhang, J.; Qian, H.; Zhang, Y.; Wu, H.; Bai, Y.; et al. Study of conduction and switching mechanisms in $Al/AlO_x/WO_x/W$ resistive switching memory for multilevel applications. *Appl. Phys. Lett.* **2013**, *102*, 233502. [CrossRef]
23. Belmonte, A.; Kim, W.; Chan, B.T.; Heylen, N.; Fantini, A.; Houssa, M.; Jurczak, M.; Goux, L. A thermally stable and high-performance 90-nm Al_2O_3/Cu-based 1T1R CBRAM cell. *IEEE Trans. Electron. Devices* **2013**, *60*, 3690–3695. [CrossRef]
24. Goux, L.; Opsomer, K.; Degraeve, R.; Müller, R.; Detavernier, C.; Wouters, D.J.; Jurczak, M.; Altimime, L.; Kittl, J.A. Influence of the Cu-Te composition and microstructure on the resistive switching of $Cu\text{-}Te/Al_2O_3/Si$ cells. *Appl. Phys. Lett.* **2011**, *99*, 053502. [CrossRef]
25. Lim, S.; Lee, S.; Woo, J.; Lee, D.; Prakash, A.; Hwang, H. Effects of N-Doped GeSbTe Buffer Layer on Switching Characteristics of Cu/Al_2O_3-Based CBRAM. *ECS Solid State Lett.* **2015**, *4*, Q25–Q28. [CrossRef]

26. Deuermeier, J.; Bayer, T.J.M.; Yanagi, H.; Kiazadeh, A.; Martins, R.; Klein, A.; Fortunato, E. Substrate reactivity as the origin of Fermi level pinning at the Cu_2O/ALD-Al_2O_3 interface. *Mater. Res. Express* **2016**, *3*, 046404. [CrossRef]
27. Rocha, P.R.F.; Kiazadeh, A.; De Leeuw, D.M.; Meskers, S.C.J.; Verbakel, F.; Taylor, D.M.; Gomes, H.L. The role of internal structure in the anomalous switching dynamics of metal-oxide/polymer resistive random access memories. *J. Appl. Phys.* **2013**, *113*, 134504. [CrossRef]
28. Wu, E.Y.; Aitken, J.; Nowak, E.; Vayshenker, A.; Varekamp, P.; Hueckel, G.; McKenna, J.; Harmon, D.; Han, L.-K.; Montrose, C.; et al. Voltage-dependent voltage-acceleration of oxide breakdown for ultra-thin oxides. In Proceedings of the 30th International Electron Devices Meeting, San Francisco, CA, USA, 10–13 December 2000; pp. 541–544.
29. Valov, I.; Tsuruoka, T. Effects of moisture and redox reactions in VCM and ECM resistive switching memories. *J. Phys. D Appl. Phys.* **2018**, *51*, 413001. [CrossRef]
30. Grzesik, Z.; Migdalska, M.; Mrowec, S. Chemical diffusion in non-stoichiometric cuprous oxide. *J. Phys. Chem. Solids* **2008**, *69*, 928–933. [CrossRef]
31. Porat, O.; Riess, I. Defect chemistry of $Cu_{2-y}O$ at elevated-temperatures. Part II: Electrical Conductivity, Thermoelectric Power and Charged Point Defects. *Solid State Ionics* **1995**, *81*, 29–41. [CrossRef]
32. Brattain, W.H. The Copper Oxide Rectifier. *Rev. Mod. Phys.* **1951**, *23*, 203–212. [CrossRef]
33. Meyer, B.K.; Polity, A.; Reppin, D.; Becker, M.; Hering, P.; Klar, P.J.; Sander, T.; Reindl, C.; Benz, J.; Eickhoff, M.; et al. Binary copper oxide semiconductors: From materials towards devices. *Phys. Status Solidi B* **2012**, *249*, 1487–1509. [CrossRef]
34. Morasch, J.; Wardenga, H.F.; Jaegermann, W.; Klein, A. Influence of grain boundaries and interfaces on the electronic structure of polycrystalline CuO thin films. *Phys. Status Solidi* **2016**, *213*, 1615–1624. [CrossRef]
35. Huang, Q.; Lilley, C.M.; Bode, M.; Divan, R.S. Electrical Properties of Cu Nanowires. In Proceedings of the 8th International Conference on Nanotechnology, Arlington, TX, USA, 18–21 August 2008; pp. 549–552.
36. Cappella, A.; Battaglia, J.-L.; Schick, V.; Kusiak, A.; Lamperti, A.; Wiemer, C.; Hay, B. High Temperature Thermal Conductivity of Amorphous Al_2O_3 Thin Films Grown by Low Temperature ALD. *Adv. Eng. Mater.* **2013**, *15*, 1046–1050. [CrossRef]
37. Timm, H.; Janek, J. On the Soret effect in binary nonstoichiometric oxides-kinetic demixing of cuprite in a temperature gradient. *Solid State Ionics* **2005**, *176*, 1131–1143. [CrossRef]
38. Kumar, D.; Aluguri, R.; Chand, U.; Tseng, T.-Y. Enhancement of resistive switching properties in nitride based CBRAM device by inserting an Al_2O_3 thin layer. *Appl. Phys. Lett.* **2017**, *110*, 203102. [CrossRef]